CITIES OF LIGHT AND HEAT

Cities of Light and Heat

Domesticating Gas and Electricity in Urban America

Mark H. Rose

The Pennsylvania State University Press
University Park, Pennsylvania

Library of Congress Cataloging-in-Publication Data

Rose, Mark H., 1942–
Cities of light and heat : domesticating gas and electricity in urban America / Mark H. Rose.
p. cm.
Includes bibliographical references and index.
ISBN 0-271-01349-4
1. Municipal services—Colorado—Denver—History. 2. Municipal services—Missouri—Kansas City—History. 3. Public utilities—Colorado—Denver—History. 4. Public utilities—Missouri—Kansas City—History. 5. Electric utilities—Colorado—Denver—History. 6. Electric utilities—Missouri—Kansas City—History. 7. Gas industry—Colorado—Denver—History. 8. Gas industry—Missouri—Kansas City—History. I. Title.
HD4606.D33R67 1995
363.6′0978139—dc20 94-16202
CIP

Printed in the United States of America

Published by The Pennsylvania State University Press,
University Park, PA 16802-1003

It is the policy of The Pennsylvania State University Press to use acid-free paper for the first printing of all clothbound books. Publications on uncoated stock satisfy the minimum requirements of American National Standard for Information Sciences—Permanence of Paper for Printed Library Materials, ANSI Z39.48–1984.

To Marsha Lynn and our daughters,
Amy Claire and Liana Isa

To Berte and S. Albert Rose, my parents
and in memory of
Ethel T. and Leo M. Shapiro

CONTENTS

LIST OF ILLUSTRATIONS

LIST OF TABLES

PREFACE

Context and colleagues matter a great deal. During the mid-1970s, I held an appointment as visiting assistant professor of history at the University of Kansas. One day, John G. Clark told me that a federal agency (with which I was unfamiliar) was awarding grants to support full-time research in a field labeled energy history. We would apply together. John had written a book on natural gas and economic development in southeastern Kansas before 1930, and I was completing revisions on a book dealing with the origins of the Interstate Highway System. As we prepared our grant application for submission to the Office of the Historian at the U.S. Energy Research and Development Administration, we noticed that Americans consumed more energy to heat their homes and drive their automobiles than their counterparts in other nations. In that period of energy shortfalls, we assumed that economic abundance and a penchant for privacy had probably shaped the American appetite for immense quantities of gasoline, natural gas, and electricity. We were also convinced that Americans had encouraged political leaders to provide them with lots of cheap energy. In order to facilitate our work, John concentrated on federal energy policy and I accepted responsibility for case studies at the local and urban level. Kansas City and Denver offered logistical advantages and the potential, I believed, for a comparative study.

We were awarded the grant. In 1977, John went to Washington, D.C., for the first of two years as visiting scholar at the U.S. Department of Energy. I remained at Kansas University as his replacement for one year. In an essay that we jointly prepared for the *Journal of Urban History,* John and I focused on consumer choices within frameworks set by American values, new technologies, and aggressive developers in the energy field.

During the next fifteen years, contexts and colleagues suggested additional dimensions to my research and writing. In 1978, I accepted a position as research associate at the Franklin Institute in Philadelphia. W. Bernard Carlson introduced me to the Society for the History of Technology and to the weekly seminar in the Department of the History and Sociology of Science at the University of Pennsylvania. Bernie and I spoke regularly about developments in urban, social, political, and technological history and how we might create integrative concepts.

I spent the 1980s in a science and technology studies program located at Michigan Technological University. George H. Daniels, a colleague, insisted that societies shaped their own technologies. Bernie Carlson, for a time also a member of the program at Michigan Tech, reinforced George's work with cases drawn from his study of inventors such as Thomas A. Edison and Elihu Thomson. By the mid-1980s, Bruce E. Seely and Larry D. Lankton were encouraging me to perceive differences between the popular meanings of science and technology. In addition, Bruce recognized that people remain the best carriers of new ideas. As General Editor of a monographic series for Temple University Press, I was also reading a number of manuscripts that were especially informative about urban politics, experts, systems, culture, and the meaning of gender for the design of machines, houses, and suburbs. Finally, I decided that books and articles by Alfred D. Chandler Jr., Ruth Schwartz Cowan, Theodore A. Hershberg, Thomas P. Hughes, Kenneth T. Jackson, and Joel A. Tarr would form the perimeters of this study.

Colleagues and context had encouraged me to alter the direction of several articles and this book. Now, I sought to explain the paths by which urban Americans defined and adapted gas and electric appliances for their homes and public buildings. For the period up to 1900, I was struck by the importance of the local political economy in determining the organization, finance, and competitive strategies of gas and electric operators. Persons long immersed in securing railroad connections and speculating in real estate, banking, and water companies identified gas and electric service as an extension of those activities. For the period after 1900, I found that gas and electric operators such as Henry L. Doherty in Denver fixed rates and trained personnel with a view toward serving diverse householders who were increasingly located in distant areas of the city. By way of contrast, not until the 1920s did operators of gas and electric firms in Kansas City create rates and levels of service that satisfied householders, especially savvy and politically sophisticated householders residing along the city's periphery.

A group I identify as agents of diffusion also played a crucial role in defining household appliances and in encouraging urban householders to purchase them. There were none more active as agents of diffusion than educators, home builders, architects, and sales personnel for gas and electric companies. These enthusiasts for gas and electric appliances spoke of comfort, convenience, and cleanliness, especially for women. Agents of diffusion instructed women in the intricacies of gas cooking and electric ironing. At the same time, men learned the details of electric wiring, and they were informed in unambiguous terms of their responsibility to provide a built environment for women that was warm, well-ventilated, odor-free, and likely to protect them from disease. During the interwar decades, so successful were these sales promotions that operators of gas and electric firms undertook substantial programs of reorganization and employee education. For the period up to 1940 in Kansas City and Denver, then, I find that political economy, urban decentralization, organizational innovation, and the remarkable effects of teachers, home builders, and others had largely determined the meaning of gas and electric service and the patterning of their adoption across urban space.

Context and colleagues encouraged me to push this thesis into the postwar period, seeking to explain patterns of diffusion for the United States as a whole. Upon reaching urban South Florida in the summer of 1990, I could not help but be struck by the extent to which many residents assumed that the built and even the natural environments ought to be controlled down to the smallest detail. Wealthy Floridians often swim in heated pools during the summer!

The ability of Americans to swim in heated pools extended from their backyards to a series of institutional arrangements on the national (and international) levels that were undergoing substantial change. For about two decades following World War II, standards of comfort, convenience, and cleanliness fixed before 1940 encouraged a rapid diffusion of appliances. In order to supply this growing demand, engineers at the nation's gas and electric firms increased output, reduced costs, and delivered service to householders at even greater distances from generating plants. During that same time, a new generation of educators, home builders, architects, and salespersons focused on educating Americans, and especially women, about the advantages of this new equipment. In turn, millions took advantage of this convergence of technologies and tastes to regulate their environments from just after birth nearly up to death, and scholars continued to debate whether technology or culture was triumphant.

The postwar boom in energy consumption and the rapid diffusion of machines to regulate the environment had not been the product of any particular triumph, whether of culture or of technology. Beginning around 1970, the period of rapid increases in demand for gas and electricity came to an end. The ability of most Americans to regulate work and domestic environments was actually becoming more tentative than a few decades earlier. Rising prices forced conservation on some, leading to rooms lit less brilliantly and no longer heated and cooled to ideal levels. Changes in the regulatory environment, and the inability of engineers to boost production and once again cut prices, only added to the growing disjuncture between political economy, technical capabilities, and urban regions characterized by their light and heat. During the twentieth century, about the only constant on the gas and electric scene was the insistence of agents of diffusion that household appliances and the comfort and convenience associated with them were especially appropriate for women.

ACKNOWLEDGMENTS

Context and colleagues are especially important in nurturing the system that supports research and writing. Several chairs and deans with whom I have worked, including Manley L. Boss, Larry D. Lankton, Mary G. McBride, and John H. Winslow, performed those unremarkable acts that are crucial to scholarship. They arranged for duplication, computers, travel, space, time off, and quiet.

Historians are always in the debt of the organizations that fund travel expenses and that provide time without teaching and other responsibilities. At Michigan Technological University, I received financial support from the Dean of the College of Sciences and Arts; from the Director of the Office of Research; and from the Program in Science, Technology, and Society with funds granted by the Sohio Corporation. Outside of Michigan Tech, I enjoyed support provided by the Albert J. Beveridge Award Committee of the American Historical Association; the Herbert Hoover Library Association; the Harry S. Truman Library Association; and the Eleanor Roosevelt Institute. Particularly valuable in allowing me to take extended leave for periods of research and writing were two awards made by Dr. Ronald Overmann at the National Science Foundation. Without Ron's enthusiastic support for the social history of technology, this book would not have been written. Naturally, any opinions, findings, conclusions, and recommendations expressed in this book are my own and do not necessarily reflect the views of those who supported this research.

Managers of corporate records were generous with their time and that of their staff members. W. D. "Bill" Fitzmaurice arranged for me to spend several weeks in the records center of the Public Service Company of Colorado. In Kansas City, I was allowed access to portions of the printed records of the Gas Service Company and the Kansas City Power & Light Company. Finally, Myron D. Calkins, Director of Kansas City's Department of Public Works, arranged for me to review materials still in the possession of several city departments. Both at city hall and at each of the gas and electric firms, moreover, managers allowed me to use their copiers without charge.

Archivists and librarians were unfailingly helpful at each stage in the research process. I worked in the Kansas City Public Library, the Kansas City Museum, the Kansas Historical Society in Topeka, and in the library of the University of Kansas in Lawrence. In particular, my visits to Kansas City were made more productive by the excellence of the collections at the University of Missouri at Kansas City and by David Boutros's superb knowledge of local materials. In Denver, I made good use of the collections at the Colorado Historical Society, the Denver Public Library, and the library of the University of Denver. Helen Samuels in Archives and Special Collections at the Massachusetts Institute of Technology identified and mailed several valuable items. During the course of my residence in Philadelphia and several return visits, I benefited from the collections at the University of Pennsylvania Libraries, the library of Temple University, and the once-great collection of engineering publications formerly housed at the Franklin Institute. I have also enjoyed access to collections at Florida Atlantic and Michigan Technological Universities. Because my research shifted from the national to the local and urban scenes, I had to set aside portions of the material I had collected at the presidential libraries of Herbert Hoover, Franklin D. Roosevelt, and Harry S. Truman. I wanted, nonetheless, to recognize those first-rate collections and the knowledgeable, congenial, and overworked archivists who maintain them.

During these years of budgetary stringency, scholars have become increasingly dependent on their interlibrary loan officers. In the early stages of this book, Margaret J. Koski and Barbara J. Wilder at Michigan Technological University serviced numerous requests. At Florida Atlantic University, I was equally well treated by Sandra L. Bland, Kenneth M. Frankel, Usha Sudhakar, and Nancy H. Wynen, head of the department.

Mark S. Foster and Gregory Black rented their homes to us at ridiculously low prices, which allowed me to extend visits to Kansas City and Denver. Mark also walked, jogged, and drove me through Denver's neighborhoods, allowing me to develop a visual sense for the historic social and economic geography of a city that maps, photographs, and reports never fully provide. Thomas Parke Hughes arranged for me to use his study carrel and secured library privileges. Both Tom and W. Bernard Carlson, his graduate student and now a historian at the University of Virginia, made a political and urban historian feel comfortable at those wonderful seminars held each Monday afternoon in the Department of History and Sociology of Science at the University of Pennsylvania.

The gap between early drafts and a completed manuscript is often immense. Several persons generously read portions of the manuscript and offered a number of valuable suggestions for improvement. I am pleased to acknowledge the insights of John C. Burnham, Harvey M. Choldin, Stephen D. Engle, Nina Lerman, Margaret S. Marsh, Raymond A. Mohl, Mary Corbin Sies, Bayla Singer, and Elizabeth Toon. Still others read the manuscript in full, and I am equally pleased to recognize the perceptive comments provided by David Boutros, Mona Domosh, Claude S. Fischer, Kenneth W. Goings, JoAnne Goldman, Carolyn Goldstein, Sandra L. Norman, Harold L. Platt, Allen J. Share, David Sicilia, and William H. Wilson.

Peter J. Potter acquired this manuscript for the Penn State Press. He had confidence in a project not yet completed and not fully formed. He also provided the "hard" reading that authors always claim they want. Peter's thoughtful (and always positive) remarks added substantially to the manuscript's conceptual clarity. At the next stage, Peggy Hoover provided a superb reading as copy editor and then guided the manuscript through production.

Colleagues helped shape this book unknowingly. During the past twenty-five years, they have generously offered valuable assessments of the patterning of American history and historiography. I am particularly in the debt of Paul Barrett, John C. Burnham, W. Bernard Carlson, John G. Clark, Donald T. Critchlow, George H. Daniels, Mark S. Foster, Kenneth W. Goings, Ellis W. Hawley, Melvin Kranzberg, Donald R. McCoy, Raymond A. Mohl, Bruce E. Seely, Mary Corbin Sies, Robert A. Slayton, Joel A. Tarr, Rudi Volti, Leonard S. Wallock, Rosalind H. Williams, and especially my friend and adviser K. Austin Kerr.

During the early stages of research and writing, our daughters, Amy Claire Shapiro Rose and Liana Isa Shapiro Rose, waited up late for their father to get home from research trips. As they reached their teen years, I learned a great deal from them about gender and the built environment. Amy and Lolly now add immeasurably to the pleasure of being middle-aged. My wife, Marsha Lynn Shapiro Rose, remains busy with her career as a sociologist. Marsha Lynn still creates the intellectual and social contexts within which I have prepared each article and book.

Introduction

IN THE SUMMER OF 1878, a traveling circus brought electric lighting to Kansas City, Missouri. Promoters of the circus boasted of "eighteen electric light chandeliers, yielding a volume of light equal to 35,000 gas jets." This device was so fabulous, they added, that the "dreary night [was] made bright and beautiful." Because electric lighting had not been seen before, continued the promoters, "people [were] coming on grand excursion trains hundreds of miles from all parts of the country." That "philosophers are bewildered . . . [by] it," continued the advertisement, and that "all scientists gaze in wonder at it" spoke as well to the magic and mystery associated with electric technology.[1]

Three years later, the sight of electric lighting continued to thrill the people of Kansas City. During the evening of March 23, 1881, operators of the G. Y. Smith Dry Goods Company turned on sixteen carbon-arc lights, the first located indoors in the city. According to a local journalist, "almost every inch of available space in street, sidewalk and even gutter was covered by a jostling, crowding, pushing, hurrying mass of people," all seeking to look at the lights. Compared with the brilliant illumination provided by the arc lights, the store's gas lamps appeared "yellow, ghastly, and ashamed of themselves." The new arc lamps, observed the journalist, represented "the splendid triumph of science." According to historian David E. Nye, the sense of thrill and excitement that Americans experienced on streets and in shops throughout the United States grew from their conviction that light was an "exemplification of Christianity, science, and progress."[2]

1. *Kansas City Star,* June 4, 1950, in Native Sons Archives, vol. 59, Kansas City Public Library, Missouri Valley Room (cited hereafter as Kansas City Public Library).

2. *Kansas City Evening Star,* March 24, 1881, clipping files, Kansas City Public Library; David E. Nye, *Electrifying America: Social Meanings of a New Technology, 1880–1940* (Cambridge: MIT Press, 1990), p. 36. For the concepts of luxury and novelty in early electric lighting, see Carolyn Marvin, *When Old Technologies Were New: Thinking About Electric Communication in the Late Nineteenth Century* (New York: Oxford University Press,

The introduction of electric and gas lighting in North America was part of a more general interest in the development of technological systems for growing cities. During the last two decades of the nineteenth century, business executives, city leaders, and engineers in many parts of the world began to direct installation of trolley, water, sewer, and telephone as well as gas and electric networks. Equally important was the immense and costly process of street paving. According to historians Christopher Armstrong and H. V. Nelles, construction of these urban technologies, including electric and gas technologies, was part of an institution they describe as "the theatre of science." The products of that theater, such as electric lighting, represented "the vaulting ambitions and subsequently the hallmarks of bourgeois civilization."[3]

In subsequent years, items such as electric lights and trolley systems entered the realm of the ordinary for millions of Americans. By the late 1920s, electric lighting, however modest, was a standard feature in the homes of most urban Americans. During the period after World War II, the attention of householders turned to a new generation of electric and gas appliances such as television sets, forced-air furnaces, and electric garbage disposals. By the 1960s, as those items became commonplace, Americans began to purchase another generation of exciting equipment for their houses, including air conditioning.

At first glance, the widespread distribution of electric lights, forced-air furnaces, or air conditioning followed a similar and nearly certain path. Scientists, the leading actors in the theater of science, discovered and built early versions of products, which were then converted by engineers, mechanics, and manufacturers into useful and inexpensive appliances for the millions. The inherent superiority of the new product guaranteed that it would soon replace the older one. By the 1920s, as this account would have it, bright and reliable electric lights had largely replaced dull and often dangerous gas lights. Following World War II, forced-air furnaces burning natural gas replaced coal stoves and furnaces. Out went the soot and smoke long associated with coal.

1988), pp. 158–190; and Mark J. Bouman, "Luxury and Control: The Urbanity of Street Lighting in Nineteenth-Century Cities," *Journal of Urban History* 14 (November 1987), 7–37, which is only a portion of a larger manuscript the author generously shared with me.

3. Christopher Armstrong and H. V. Nelles, *Monopoly's Moment: The Organization and Regulation of Canadian Utilities, 1830–1930* (Philadelphia: Temple University Press, 1986), pp. 11, 59.

Corporate publicists were quick to connect the enhanced comfort and convenience of gas and electric appliances with ideas such as science, progress, or even democracy. In 1913, the General Electric Company advertised fans and other small appliances as providing "a luxury once reserved for the rich and now made universal by electricity." During the 1930s, the nation's largest corporations sponsored exhibits at fairs held in Chicago and New York. Designers of those exhibits stressed the comfort and convenience that their companies' products would soon bring near-at-hand. Altogether, popular writers depicted the technological past as an introduction to the present. "The message" at these fairs, reports historian Robert W. Rydell, was "that science was modern man's salvation and that the scientist engineer was priest—if not savior."[4]

Historians have long recognized that scientists and engineers were not the all-powerful agents depicted by their early promoters. For example, Nelles and Armstrong, in their analysis of the development of gas, electric, trolley, and other elements of the "vaulting ambitions" of middle-class Canadians, determine that urban and provincial politics influenced the size of gas and electric systems, who controlled service, who received service and when, and the rates they were charged. If politics mattered in the history of technology, so did gender. Ruth Schwartz Cowan has demonstrated that appliances reduced the most difficult and burdensome tasks of household management, such as cleaning rugs and cleaning clothes. Yet the net result of bringing vacuum cleaners and washing machines and other appliances into the home was to release men from any household responsibilities and to increase the number of tasks assigned to women.[5] Social and political contexts, then, shaped the pace and the direction by which North Americans adopted new appliances.

Context is one of the core ideas of contemporary historical scholarship, including the social history of technology and cities. In this study of gas and electric systems and of increasing demand for appliances and service, however, I find that historians have overlooked a number of contexts. One of those contexts was cities, including their dynamic politics, rapid population increases, and fast-growing suburban dis-

4. "A Luxury Once Reserved for the Rich and Now Made Universal by Electricity," advertisement, *Electrical Merchandising and Selling Electricity* 12 (June 1913), 169; Robert W. Rydell, "The Fan Dance of Science: American World's Fairs and the Great Depression," *ISIS* 76 (December 1985), 531.

5. Armstrong and Nelles, *Monopoly's Moment;* Ruth Schwartz Cowan, *More Work for Mother: The Ironies of Household Technology from the Open Hearth to the Microwave* (New York: Basic Books, 1983).

tricts. Equally important in creating contexts for gas and electric service were educators, home builders, architects, and the executives and salespersons who worked for the great gas and electric corporations. I describe those educators, executives, and others as agents of technological diffusion. Those agents defined gas and electricity as part of the effort of Americans to enhance comfort and convenience; and nearly as often, agents of technological diffusion defined appliances as appropriate for men or women, but never for both.

In the middle of the tumultuous growth that has characterized urban America since the introduction of electric lighting, neither politician nor teacher or anyone was able to assign a period of time to any particular context. As participants in a moment of remarkable innovation, agents of diffusion such as educators recognized that social, urban, and technological change were occurring in a disorderly fashion. Even so, persons who founded gas and electric firms, sold stoves, or educated children focused on politics in one period, on intense educational efforts in another, and on organizational innovation in yet another period. It was a matter of emphasis. In the next section of this introduction, I want to recreate those contexts on an individual basis and suggest a useful periodization. In the closing section, I want to explain why I studied Kansas City, Missouri, and Denver, Colorado, up to 1940 and the United States as a whole for the decades after World War II.

Up to 1900, the most important context in shaping local gas and electric operations was the political economy of these two booming cities in the American Midwest. Even before gas and electricity were available, the workings of that political economy had encouraged business and political leaders to conceive of the growth of their cities and the growth of their own private wealth in terms of external and internal actors. In their dealings with outsiders, the prevailing wisdom was that unity was the wisest strategy. Business leaders, publicists, and politicians regularly celebrated the prospect of securing railroad connections to the East. Rare, in fact, was the western urbanite who doubted the advantages for local business and property values of eastern capital and a station on the main line to the finances and commerce of Chicago and New York. Relative to the larger cities of the East, the political economies of Kansas City and Denver were not simply permeable, they were porous.

Cooperation vanished on the local scene. Savvy business leaders created rival and shifting groups of investor-managers. Members of these groups competed fiercely with one another for a share of the immense wealth to be earned in providing the physical and social infrastructure of these cities and the daily needs of a growing population. Likely areas of investment and competition included meat packing, newspaper publishing, and banking. As populations doubled and then doubled again, competing investors poured resources into combined real estate and trolley ventures. Success in many of these undertakings required the approval of local politicians, especially for enterprises using public rights-of-way such as water, railroad, and trolley companies. That politicians were themselves investors in syndicates competing for public favor hardly seemed unusual to those who had spent their adult lives in organizing those groups and betting their own fortunes and the fortunes of others on businesses and technologies that were generally unknown, obviously risky, and potentially lucrative.

The electric and gas companies founded by these groups and approved by those local politicians rested on an identical set of perceptions and arrangements. Residents and leaders in these western cities identified gas and electricity as new players in the prestigious "theatre of science." No need to try to slow their arrival. Indeed, the burden of proof would have rested on those who were foolish enough to doubt the tangible benefits to be had from the arrival of outsiders bringing patents, machinery, and know-how.[6]

Once these new technologies actually reached the city, competition emerged among groups of investors and managers. The trick for retailers, politicians, realtors, and bankers who were excited about the prestige and commercial potential of gas and electricity was to create a syndicate, win the favor of cooperative politicians, and then secure a sufficient number of customers to cover the considerable expenses of making extensions. In a situation in which politicians were anxious to bestow new franchises on those bringing competing technologies and fresh batches of capital, neither arriving first nor possession of a superior technology guaranteed survival in the wars of rates, service, and extensions that followed.

Other than the necessity of locating near customers and railroad lines, no technical imperatives guided operators of these early gas and electric firms. Directors of gas and electric companies in Kansas City and Denver had to build a customer base, make costly extensions to the urban periphery, reassure investors, negotiate rates, and keep an eye fixed on politicians, competitors, and evolving technologies. In short, they had to succeed within the frameworks established by every-

6. Charles N. Glaab and A. Theodore Brown, *A History of Urban America* (New York: Macmillan, 1967), pp. 112–129, 147–153, 163–164, 184–187.

day commercial and political practice. Those accustomed to participating in the political economy of the late nineteenth-century city and nation had no choice but to define gas and electric businesses along identical lines.

The politics and economics of cities such as Denver and Kansas City were never static. By the mid-1890s, a desire to smite large corporations was emerging as a routine feature in local and national deliberations about trolley, water, gas, and electric franchises. In virtually every city, the talk among formerly cooperative politicians and even business executives was of regulating rates; and in many cities, conversations turned to municipal ownership. During 1896, moreover, a rate war between gas and electric firms in Kansas City forced prices well below costs. Adding to the confusion and uncertainty surrounding gas and electric operators in both cities were the immense costs of upgrading equipment to provide the service that politicians and consumers were demanding, and doing so at rates they found acceptable. By the late 1890s, then, members of the original groups of gas and electric investors in Kansas City and Denver identified gas and electric firms less as promising arenas for speculative maneuvering and more as a series of political and economic problems. By 1900, most of the original investors had vacated their positions in favor of outside capital, citywide technologies, and managers experienced in the new politics of fast-growing cities.

Around 1900, executives employed by national holding companies assumed direction of local gas and electric companies. Henry L. Doherty emerged as the principal executive of a combined gas and electric firm in Denver. In Kansas City, R. E. Richardson directed electrical operations for J. Ogden Armour, a member of the packing family who also owned the local trolley company. Natural-gas supplies in Kansas City remained in the hands of managers who took their instructions from the gas combine in Philadelphia. Altogether, then, executives on the local scene such as Doherty and Richardson brought with them the emerging tools of corporate management of technologically oriented firms. Even in remote places such as Kansas City and Denver, new managers created complex organizations directed by specialists who, in turn, spent immense sums on marketing efforts and on up-to-date distribution systems.

Imposition of the new tools of management and the expenditure of millions of dollars made little difference in determining the success or failure of gas and electric managers. An orientation toward the political geography and public policies of their cities counted far more. In 1901, with the award of a franchise to a rival electric company and the beginning of a rate war, Henry Doherty fixed a rate structure and a service plan that suited the wishes of higher-income householders. They wanted light and heat around the clock and no penalties attached to their rates for living in peripheral locations miles from electric generating plants and gas storage tanks. By 1905, moreover, Doherty's staff of more than forty salespersons had developed an appreciation for the diverse needs of saloon-keepers, recent immigrants, business executives, and wealthy housewives located in distant and increasingly fashionable Park Hill. In 1906, Denver voters awarded a twenty-year franchise, taking Doherty's firm "out of politics." By way of contrast, not until the 1920s (following periods of bankruptcy and receivership) did new operators of gas and electric firms in Kansas City learn how to adapt their rates, organizations, technologies, marketing efforts, and educational programs to the politics of a growing and diversifying city.

Doherty went further. In 1910, he named his national holding company the Cities Service Company. That name symbolized creation of an organization capable of recognizing changes in land use and politics in Denver and in many other cities, and linking those changes to organizational and technical innovations. Doherty's executives and their large staffs also brought routine to the process of identifying the changing household locations of rich and poor and the evolving tastes of black and white, male and female. Before World War I, urban growth and politics had permeated to the core of gas and electric firms managed by Doherty and by other successful executives, such as Samuel Insull in Chicago.[7]

As part of their adaptation to the city, Doherty and members of his organization also took leading positions in the task of informing householders about the advantages of electric lighting, electric irons, gas stoves, and gas-fired hot-water heaters. At first, Doherty and his employees lacked experience in selling these new appliances, but soon learned that focusing on several common concerns of the day—cleanliness, comfort, and convenience, especially for women—made sales. Husbands learned that they were responsible for reducing discomfort for mothers and wives still required to cook meals over a coal-fired stove and then clean away the soot left behind on draperies and window ledges. In short, members of the first group of gas and electric salespersons helped construct the meaning and uses of several

7. Thomas P. Hughes, *Networks of Power: Electrification in Western Society, 1880–1930* (Baltimore: Johns Hopkins University Press, 1983), p. 18.

new appliances, and in a fashion that produced solid returns on investment.

After 1900, teachers, architects, and home builders also began to instruct urban residents in the uses of gas and electric appliances. By 1915, educators informed students about the importance for sight of bright lights without glare and the importance for digestion of cooking foods on a gas stove. And at vocational schools in Kansas City, Denver, and other cities, young men not lucky enough to purchase these expensive machines learned the details of installing and repairing them, while young women learned the details of gas cooking and electric sewing.

Before World War I, however, only a small percentage of the residents of Kansas City and Denver, or any other city, actually lit their homes with electricity and cooked with gas. These earliest agents of diffusion had prepared Americans for the massive sales campaigns that followed the war. By 1920, as those campaigns got under way, educators and salespersons could build on the idea that women handled most of the new household equipment, and the same agents of diffusion could build on the increasingly popular idea that enhanced levels of cleanliness, comfort, and convenience could be achieved through mechanical means.

During the interwar decades, the sale of gas and electric appliances in Kansas City and Denver achieved regular form in the hands of such home builders as J. C. Nichols, developer of Kansas City's fabulous Country Club District. Nichols and his counterparts in virtually every urban area built homes for the well-to-do near shopping, trolley lines, and boulevards. Nichols and others like him not only brought an element of predictability to home buying, but also fostered another decade of outward movement that after 1945 would lead to suburbanization of members of the middle and working classes. During the interwar decades, however, Nichols build all-electric houses as novelty items, and constructed block after block of homes featuring gas furnaces, electric refrigerators, and gas stoves.

Nichols's Country Club Plaza, the nation's first shopping center, served women in another way. A day on the lovely plaza, with its mix of retailers, professional offices, restaurants, and free parking, encouraged the idea that shopping was a form of recreation and another opportunity to demonstrate devotion to family. Department stores located downtown were also promoting shopping as a frequent and pleasurable activity for women. What made the plaza different was its bright lights, abundant parking, refrigerated foods, home delivery, and less coal dust and smoke. Whether on the plaza or in the nearby homes of the Country Club District, Nichols was assembling the technologies of comfort and convenience for the city's most affluent families. He was also adding to the emerging gender identity of those technologies.

Roy G. Munroe was a less visible agent of gas and electric diffusion during the 1920s and 1930s. In his post as gas new-business manager at Denver's gas and electric company, Munroe never enjoyed the discretion and recognition reserved for such prominent executives as Nichols. Decisions about prices, products, and new promotions were made by his superiors in New York and Denver. Munroe's assignment was to direct the sale of appliances that others had chosen for him. Munroe served as an intermediary between policy choices and engineering expertise on the one side, and a generally untrained demand for comfort and convenience on the other. Within that circumscribed realm, he assembled a large group of representatives and trained them in the sale of gas furnaces. Naturally, he invoked his own experiences, encouraging members of his salesforce to sell furnaces in terms of the guilt that men in comfortable offices must feel as wives and mothers stoked coal furnaces and cleaned soot. By the mid-1930s, as senior executives transferred him to another assignment, Munroe took pride in the fact that nearly 20 percent of the city's householders heated with gas.

The results of efforts by Munroe and Nichols to foster gas and electric consumption extended beyond the introduction of countless appliances and instruction to sales personnel and residents of their cities in their uses and meaning. Nichols offered himself and his staff as experts in the aesthetics and technologies prerequisite to protecting families in a time of change, while he constructed an enclave for the wealthy that emphasized gas and electricity as among the most important elements in that defense. Munroe and his counterparts in Kansas City extended the skirmish line outward to include persons of more modest circumstances. Even so, every large city like Kansas City and Denver included neighborhoods (usually located nearer downtown) in which less-fortunate residents heated and cooked with coal, made do with only a few lights, cooled with ice, and picked up dust with brooms and rags. One of the indirect results of the efforts of Munroe and Nichols was to foster creation of a few neighborhoods in which residents spoke confidently about shaping an environment with the aid of air conditioning, electric lights, and gas heat. In most other sections of their cities, however, residents continued to watch the price of coal and pass information from one generation to the next about starting a cranky coal stove on chilly mornings.

For a period of about twenty years after World War II, many of the technological and social themes developed during the first half of the century converged in the form of massive urban growth and remarkable increases in the consumption of light and heat. Nothing was automatic about this process. Instead, home builders (such as J. C. Nichols) constructed houses and apartments by the hundreds or thousands (for the well-off as well as for those possessing more limited resources). State and federal officials financed great express highways, encouraging construction of homes, offices, and immense shopping malls in areas still farther from downtown. In turn, operators of gas and electric firms boosted capacity and extended service areas. Following the war, yet another generation of home economists, home builders, architects, teachers, and ordinary salespersons focused their talents on marketing furnaces and appliances by attaching them to prevailing ideas about cleanliness, comfort, and convenience, and gender.

Results of those efforts were impressive. By the late 1960s, well-off persons departed from air-conditioned offices in air-cooled automobiles, reaching homes in distant suburbs in time for a swim in a heated pool, a cooled drink, and a bit of relaxation before eating a gas-cooked meal in an air-conditioned room. The convergence of technology and culture appeared nearly perfect as educators as well as political and business leaders coalesced around the idea that a large portion of the population was entitled to regulate their environment down to the smallest detail.

By the early 1970s, the period of urban, technological, and political convergence was ending. Middle-class householders continued to move farther from the central city; and rare indeed were Americans in any locale who actually debated the importance for their own comfort of abundant and inexpensive supplies of gas and electricity. Nor were there many who believed that women were not entitled to purchase the machines and appliances that provided special protection from urban and environmental hazards. Yet several of the patterns of gas and electric production and consumption that had been developed earlier in the century did not extend into the 1970s. Most important, the price of natural gas and electricity increased dramatically. In part, factors outside the control of any person or organization, such as gas shortages and inflationary pressures, were forcing those prices upward. Similar to the period around 1900, householders and business leaders demanded relief from high prices. Unlike that period, however, no quick linkage of technology and public policy appeared in sight.

Experience had proven to be a poor teacher. During the period between 1895 and 1914, urban politicians had initiated a regulatory program that encouraged perceptive executives, such as Doherty, to launch massive programs of sales and production. Yet by the 1970s, engineers had reached the limits of several of their technologies, making it impossible, at least for the short run, to re-create the golden age of falling prices and rising consumption. In retrospect, the much celebrated convergence and momentum of technological, urban, and political systems was simply an artifact of a couple of unique circumstances stretched across a period of about twenty to thirty years. About all that remained certain was the aggressiveness of agents of diffusion in alerting Americans to new appliances, and the overwhelming desire of most Americans to achieve higher levels of cleanliness, comfort, and convenience, especially for women. But that set of ideas had been there right from the beginning.

The choice of Kansas City and Denver as the principal sites for this study demands an explanation. Several factors oriented me toward these two cities. At the start, I was impressed by the fact that Denver was among the Instant Cities described by historian Gunther P. Barth.[8] For the period up to 1870, I was also struck by the nearly identical efforts of leaders in both cities to boost property values, boast about urban greatness, and secure railroad connections. With the arrival of those rail connections, Kansas City grew more rapidly than Denver. Yet the economic activities, demography, and social and political geography of those cities remained similar up to 1940.

Kansas City and Denver were not, however, simply replicas of one another. The energy bases of the two cities was different. Kansas City was adjacent to vast supplies of natural gas; Denver, when coal was king, was surrounded by a huge supply of clean-burning lignite. But early in my research I noticed that patterns of production and consumption of hygienic, comfortable, and convenient gas and electricity in both cities converged after 1925. Despite a distance of about 500 miles, and in spite of differences in politicians, political structuring, investors, and utility managers, I could not help but be impressed by the extent to which urban politics, middle-class tastes, and the social-spatial composition of the city had triumphed.

These two cities also offered the advantages of a large corpus of research materials and a research tradition. During the 1950s, soci-

8. Gunther P. Barth, *Instant Cities: Urbanization and the Rise of San Francisco and Denver* (New York: Oxford University Press, 1975).

ologists and historians at the University of Chicago selected Kansas City as a principal research site, arguing its typicality and accessibility. In turn, historians R. Richard Wohl, Charles N. Glaab, and A. Theodore Brown selected Kansas City for detailed study of their contention that entrepreneurs set the framework for the urbanization process. Altogether, I thought, here was an opportunity to test a couple of hypotheses and to add to a body of literature extending back to the University of Chicago and to the Chicago School of Sociology. In the process of relocating gas and electricity in cities, I also perceived an opportunity to excavate the remains of earlier scholarly traditions.

After 1945, concentration on Kansas City and Denver no longer appeared useful. I do not doubt that local managers of gas and electric firms continued to attend to the social and political geographies of their cities and regions. Yet the initiative in the political arena regarding rates and service had shifted upward to state and federal agencies and to trade groups located in the largest urban centers. I also recognized that educators, home builders, or utility employees in most cities no longer controlled the initiatives or many of the details involved in bringing fresh ideas to urban residents about the uses of gas and electric appliances. During the postwar years, patterns of production, consumption, and regulation developed decades earlier in a number of cities converged in the form of rapid increases in demand for gas and electric service throughout the nation. As a research strategy, I thought, concentration on the national scene after 1945 would allow a clearer view of developments in countless American cities.

In order to comprehend these changing patterns of production, consumption, and regulation, it is necessary to begin in Kansas City and Denver during the period between 1860 and 1900. In those booming cities, persons who had founded banks, speculated in real estate, created retail stores, and developed horsecar service and water supplies determined that gas and electric service would prove another desirable outlet for their optimism and cash. The first gas and electric firms were extensions of the idea that it was possible to secure wealth and the prestige of science by linking urban growth, public policy, and new technologies in the form of gas and electric service. Founders of these firms set in motion the political economy that led eventually to creation of cities of light and heat.

CHAPTER ONE

The Urbanization of Technology and Public Policy, 1860–1900

> The Americans were very enterprising in business; they worshipped the almighty dollar, and looked upon the electric light as a good advertisement. A man would, for the sake of outshining his rival, put the most flickering arc light outside his door . . . [Yet] the man who would have it in his store would not think of using it at home to make it more comfortable, because it cost too much.
>
> —W. H. Preece
> English visitor to the United States, 1884

OPERATORS OF THE FIRST GAS AND ELECTRIC companies in rapidly growing Kansas City and Denver were members of the groups that had guided investments, growth, and politics in their cities for two or three decades. Gas and electricity involved new technologies, esoteric knowledge, and an uncertain demand. During the 1870s and 1880s, however, those long-accustomed to competition and growth in such fields as banking, real estate, railroads, and horsecars turned to gas and electric franchises as another arena in which to position themselves to benefit from public policy and rapid urban growth.

Not even directors of prosperous banks could guarantee the future of new gas and electric companies. During the 1890s, the pace of urban growth and technological innovation challenged the finances and managerial savvy of the first generation of operators. A decade earlier, politicians interested in contributing to the city's growth had granted franchises without fees or service requirements to each applicant. By the mid-1890s, however, political leaders demanded universal service to distant neighborhoods at uniform and low rates. By 1900, several of the original operators of gas and electric firms in Kansas

City and Denver had vacated the field in favor of outside interests possessing additional capital and more sophisticated understanding of new technologies. Up to that point, operators failed to adapt to rapid urban change, especially to changes in the social and political geography of the cities that they had done so much to shape.

Kansas City and Denver were founded as entrepôts for western settlement and as centers of real-estate speculation. Kansas City was launched first. In November 1838, investors located in small towns near the junction of the Missouri and Kansas rivers joined with counterparts in St. Louis to found the Town of Kansas. During the 1840s, thousands moved through the region, seeking land for farming, gold, and an opportunity to supply migrants with clothing and implements. By the early 1850s, local grocers, millers of wood and grain, wagonmakers, blacksmiths, and hotel operators provided the rudiments of urban services for farmers and travelers. In February 1853, the Missouri legislature awarded a charter to the City of Kansas. By 1860, on the eve of the Civil War, the city had a population of about 4,700.[1]

Residents of Kansas City also had developed several institutions that focused on the business of urban growth. Town government was in the hands of local promoters and publicists, who were dedicated to growth, especially the growth of real-estate prices believed certain to follow often speculative investments. Creation of "paper" railroads was another of the time-honored methods by which local speculators and investors hoped to attach their fortunes to urban growth. Even the prospect of civil war failed to reduce enthusiasm for real-estate investments or for schemes to bring railroads to the city.

Not all was image. Civic leaders in Kansas City had managed to create several streets, overcoming technical and financial challenges in the face of steep and rocky bluffs at the confluence of the two rivers. In a political economy predicated on faith in the future, residents of the city and nearby towns also benefited from federal outlays amounting to $3.2 million annually for army disbursements, annuity payments to Native Americans, and mail contracts. By 1860, Kansas City had crossed the urban threshold.[2]

1. A. Theodore Brown and Lyle W. Dorsett, *K.C.: A History of Kansas City, Missouri* (Boulder, Colo.: Pruett Publishing Company, 1978), pp. 5–6.

2. Ibid., pp. 5–23; Stuart Blumin, *The Urban Threshold: Growth and Change in a Nineteenth-Century American Community* (Chicago: University of Chicago Press, 1976), pp. 2–9. For railroads and real-estate booming in Kansas City, see Charles N. Glaab, *Kansas City and the Railroads: Community Policy in the Growth of a Regional Metropolis* (Madison: State Historical Society of Wisconsin, 1962), pp. 36–60, 171; and for rail-

Denver grew by equal measures of fortunate timing, luck, and the talent of decisive leaders. Up to the late 1850s, thousands had migrated westward in search of farms, business opportunities, and, as ever, gold. In 1858, a gold strike that occurred around Pike's Peak attracted the attention of people throughout the nation. By 1859, two small settlements had developed alongside the Cherry Creek River. Residents of one of these chose the name "Auraria" for the settlement, after a gold region in Georgia. Residents of the other settlement preferred the name "Denver," for General James W. Denver, the territorial governor of Kansas. In April 1860, members of the rival groups agreed to unite under the name "Denver," following instructions of General William Larimer. Larimer had physically occupied the small territory of the opposition land company in Auraria. His overriding concerns were those of railroad connections to the east, rising property values, land sales, and a political career for himself. In 1862, however, following disappointments in the political arena, Larimer left the city.[3]

Those who remained in Denver shared Larimer's passion for rapid growth. Only a few secured the fortunes and prominence that gold bestowed. Others earned substantial livelihoods in real-estate speculation, banking, and law, and in saloon-keeping, which was still the city's largest industry. Not all prospered, and those who did not joined a transient work-force of servants, cooks, mechanics, prostitutes, and general laborers. Between 1860 and 1870, more than 100,000 persons and $27 million in gold dug from nearby strikes passed through Denver, creating profits and jobs and lending substance to the hopes of still others that riches would soon be theirs. In 1870, Denver remained an "instant city," to borrow historian Gunther Barth's useful

roads and real-estate promotions as components in city-building throughout the Midwest, see William Cronon, *Nature's Metropolis: Chicago and the Great West* (New York: W. W. Norton & Company, 1991), pp. 32–41, 63–74. Finally, older and still valuable accounts of entrepreneurial enthusiasm, town-building, and physical improvements include Carl Abbott, *Boosters and Businessmen: Popular Economic Thought and Urban Growth in the Antebellum Middle West* (Westport, Conn.: Greenwood Press, 1981), pp. 24, 133, 198–208; and Richard C. Wade, *The Urban Frontier: Pioneer Life in Early Pittsburgh, Cincinnati, Lexington, Louisville, and St. Louis* (Chicago: University of Chicago Press, 1959), pp. 77–79.

3. Lyle W. Dorsett, *The Queen City: A History of Denver* (Boulder, Colo.: Pruett Publishing Company, 1977), pp. 1–8. For the more general context of building towns along rivers, real-estate speculation, and politics during this period, see Timothy R. Mahoney, *River Towns in the Great West: The Structure of Provincial Urbanization in the American Midwest, 1820–1870* (New York: Cambridge University Press, 1990), pp. 101–107.

construct. Like the City of Kansas a decade earlier, however, Denver had also crossed the urban threshold.[4]

Following the Civil War, leaders in both cities made the politics of gaining railroad connections preeminent. Small groups of local enthusiasts took the initiative in political and financial circles. As before, prospects of a great city and of the profits to be made from exercising options on land near the expected rights-of-way encouraged members of these syndicates to persevere. During the next few years, they assembled financial and political packages (including, especially, land grants) that railroad executives judged prerequisite to establishing service across the Missouri River. On July 3, 1869, the Hannibal and St. Joseph Bridge opened for railroad traffic across the Missouri River, assuring Kansas City's dominance in regional trade. By 1880, according to historian Lawrence H. Larsen, "Kansas City was the most important transportation hub in the nation west of Chicago."[5]

Leaders in Denver had to devise a more creative approach to securing a railroad connection than did their counterparts in Kansas City. Officials of the Union Pacific Railroad were not planning to construct a line to Denver, preferring instead the easier route westward through Cheyenne, Wyoming. Rather than allowing their moment for achieving urban greatness and hefty profits to pass north of Denver, however, local leaders raised funds through subscription and financed a link to Cheyenne. By August 1870, lines of the Union Pacific as well as the Kansas Pacific served Denver, assuring the city's future growth.[6]

With railroad connections in place, each city joined an emerging network of cities. Because Kansas City and Denver were no longer remote outposts but regular stops, those two cities emerged as routine outlets for social, technological, and industrial change. Immigrants from Europe, Asia, and the rural United States moved to Kansas City and Denver, leading to spectacular increases in population (see Table 1). Capitalists and technologists also included Denver and Kansas City as likely places to secure returns on their investments or on their specialized knowledge of trolley, water, and other technical systems that residents of many cities were becoming excited about. After 1870,

4. Gunther P. Barth, *Instant Cities: Urbanization and the Rise of San Francisco and Denver* (New York: Oxford University Press, 1975), pp. 140–141, 144, 148–154, 160–161, 175, 213.

5. Charles N. Glaab and A. Theodore Brown, *A History of Urban America* (New York: Macmillan, 1967), p. 116; Lawrence H. Larsen, *The Urban West at the End of the Frontier* (Lawrence: Regents Press of Kansas, 1978), p. 9.

6. Dorsett, *Queen City,* p. 23.

Table 1 Populations of Denver and Kansas City (Mo.), 1860–1900

	1860	1880	1900
Denver	4,749	35,629	133,859
Kansas City	4,418	55,785	163,752

SOURCE: U.S. Department of Commerce, Bureau of the Census, *Sixteenth Census of the United States: 1940,* vol. 1 (Washington, D.C., 1942), pp. 32–33.

national themes in the form of industrial growth and social change converged in Kansas City and Denver, as in every large city.[7]

During the period between 1870 and 1900, members of the founding generations in Kansas City and Denver continued to act on the convictions that had aided them in establishing their cities, making lucrative investments, and securing railroads. Politics and economics, they believed, were naturally tied to one another. The key to personal wealth and to still more impressive cities, then, was to assemble a group of investors who would move decisively to combine new technologies and public franchises. Gas and electric service offered yet another set of opportunities for their capital, entrepreneurship, and political know-how.[8]

The process of installing gas streetlights had begun in Europe and North America around the same time. In the United States, Baltimore was the first city to do so, beginning in 1816. Thereafter, political leaders in most of the larger cities, such as Boston, New York, New Orleans, and Louisville, ordered installations. Amateur governments, to borrow historian Harold L. Platt's instructive concept, contracted with private companies to supply gas and service lamps. Although wealthy households and managers of larger offices purchased gas for indoor lighting, the city, as the principal customer for street lighting,

7. Brown and Dorsett, *K.C.,* p. 37; Glaab and Brown, *History of Urban America,* pp. 114–118; Michael P. Conzen, "The American Urban System in the Nineteenth Century," in D. T. Herbert and R. J. Johnson, eds., *Geography and the Urban Environment: Progress in Research and Applications,* vol. 4 (New York: John Wiley & Sons, 1981), pp. 331–335. Barth, *Instant Cities,* pp. 208–232, contends that the arrival of transport and other technological systems converted these instant cities into ordinary ones.

8. For accounts of the politics of trolley and real-estate development in Kansas City and Denver, see Allen duPont Breck, *William Gray Evans, 1855–1924: Portrait of a Western Executive* (Denver, Colo.: University of Denver, 1964), pp. 110–113; and Theodore S. Case, ed., *History of Kansas City, Missouri* (New York: D. Mason & Company, 1888), pp. 407–417. For a perceptive account of the relationships among public works such as bridges and speculation in real-estate by members of competing syndicates in Chicago, see Robin L. Einhorn, *Property Rules: Political Economy in Chicago, 1833–1872* (Chicago: University of Chicago Press, 1991), pp. 42–60.

helped fix prices for everyone. After a while, customers started to believe that the lighting could be brighter, the service more reliable, and the costs cheaper.[9]

The gas lighting idea reached Kansas City and Denver along with the first days of growth, land booming, and railroad promotion. As early as 1858, authorities in Kansas City granted a franchise to a Goodwin, Northrup, and their associates to operate a gas works. A year later, officials in Denver awarded a franchise for gas operations to several local persons. In both cases, however, the franchise holders failed to build the plants. In July 1868, Denver's town assembly awarded a gas franchise to a Colonel Heine in Paris, France, on the basis of a letter promising construction upon receipt of the franchise. That franchise also went unfulfilled. Even to persons long-accustomed to the uncertainty of finding gold and to the intricacies and tentativeness of competing with nearby towns for bridges and railroad lines, the gas business appeared too uncertain. Gas plants, like so many elements of town building in these communities, had to await the arrival of rail connections and the long-expected growth and boom.[10]

Organizers of Denver's first truly operational gas company were leaders in the business of booming real estate and constructing the city's infrastructure. On November 13, 1869, eight men incorporated the Denver Gas Company. Each was a recent migrant to the region and city. Several had been active in establishing the city's railroad connections, while others directed banks and newspapers. During the course of a lengthy and successful career, John Evans comfortably mixed his medical training and several years in practice with subsequent work as territorial governor, management of trolley firms, real-estate promotion, and founding of two universities. Lewis N. Tappan,

9. Martin V. Melosi, *Coping with Abundance: Energy and Environment in Industrial America* (Philadelphia: Temple University Press, 1985), p. 58; Daniel J. Boorstin, *The Americans: The Democratic Experience* (New York: Vintage Books, 1974), p. 43; Harold L. Platt, *City Building in the New South: The Growth of Public Services in Houston, Texas, 1830–1910* (Philadelphia: Temple University Press, 1983), pp. 7, 21–22; Christopher Armstrong and H. V. Nelles, *Monopoly's Moment: The Organization and Regulation of Canadian Utilities, 1830–1930* (Philadelphia: Temple University Press, 1986), pp. 3–33.

10. Kansas City, Missouri. Public Utilities Commission, *Laws, Ordinances, and Permits Dealing with Rights, Privileges, and Franchises of Public Service Corporations in Kansas City, Missouri* (1912), p. 12; Public Service Company of Colorado, Employee Educational Committee, "History of the Company," 1935, typescript and mimeograph at the State Historical Society of Colorado, Denver (cited hereafter as SHSC); Dorsett, *Queen City*, p. 79.

who held the original horse-railway franchise granted in 1867, was also a member of this group. Another principal was Walter S. Cheesman, who had earned hefty profits in the real-estate and drug businesses and who was active in railroads, banking, and development the following year of the Denver City Water Company. The chief organizer of the gas project, James Archer, also was a railroad promoter, tied in particular to the Kansas Pacific, and later co-founder with Cheesman of the water company. As a reward for his efforts in organizing this project, Archer was elected president of the firm. Similar to the way real-estate and railroad promotions were being organized, a group of Denver investors had created a syndicate with a view toward technology, urban growth, and a public franchise. This time the medium for their expression was gas.[11]

City officials approved the project on the spot. On November 13, the day of incorporation, a special committee of the council reported favorably on earlier negotiations with Archer and his associates. The price of $5 per thousand cubic feet (mcf) for manufactured gas "seemed high," as members of the committee figured it, but they also were willing to take account of the proposition that "houses in Denver were few and far apart and required . . . long lines of pipes to be laid." The net result was that city officials awarded a contract to Archer's group for a period of fifty years beginning January 1, 1870. Archer postponed construction of the gas house and pipe-laying until September 1870, a month after the arrival of the first railroad.[12]

During the next five years, the preferences of the founders shaped gas rates and operations. In December 1870, with construction under way, city and company officials negotiated a rate of $50 per lamp each year for five years. The rate of $5 per mcf for household lighting remained in effect. In January 1871, the first lights were turned on, and lamp-lighters began their fabled rounds. At first, only two miles of main were installed, but eight miles were in place by the end of the year.[13] At a rate of $50 for street lamps and $5 per mcf for household

11. Charles A. Frueauff, "History of the Company Gas Properties," May 1, 1905, SHSC; Dorsett, *Queen City*, pp. 74–75, 77, 79, 85–86.

12. Dorsett, *Queen City*, p. 79; *Rocky Mountain News*, November 15, 1869, typescript in Records of the Public Service Company of Colorado (PSCC), Files of Roy G. Munroe, Denver (cited hereafter as Munroe Files); Clyde Lyndon King, *The History of the Government of Denver with Special Reference to Its Relations with Public Service Corporations* (Denver: Fisher Book Company, 1911), p. 80.

13. King, *History of the Government of Denver*, pp. 80–81; "History of the Public Service Company of Colorado" (outline, c. 1958), in Munroe Files.

lighting, this period was probably a golden one for investors in the city's gas firm.

Between 1875 and 1883, the relationship between city and company officials became less cooperative. Because the city was the principal customer for gas, city officials exercised a special advantage in rate negotiations, and thus they used their dominant position as leverage. By 1882, rates had been forced down to $40 a lamp for street lighting and $3 per mcf for 20-candlepower interior lights.[14]

Furthermore, city officials used competition to their advantage. In 1883, the city awarded a franchise to the United Gas Improvement Company of Pennsylvania on the basis of gas priced for homes at $2 per mcf and $27 for outdoor lamps. In April 1887, still another dose of competition was introduced in the form of the People's Gas Light Company, with the right to charge no more than $1.50 per mcf.[15]

The early days of gas operations in Kansas City followed a similar path. In 1866, the Missouri legislature chartered the Kansas City Gas, Light & Coke Company to build and operate a gas plant. The project was begun in April 1867, even before a rail connection to the east was assured, and completed in October. The agreed-on rate for household users was $4.50 per mcf, and the company was to provide 201 street lamps at $42 each per year. By 1877, about the same time as in Denver, city officials demanded lower rates for the city and for private customers. Now, $3 per mcf looked about right, and the company was to light 323 street lamps at the reduced rate of $32.50 a year. Meetings were held and threats were exchanged between city and company officials. Company leaders introduced caprice into the bargaining, turning lights off without warning. Residents carried lanterns. By 1880, though, census officials counted 605 streetlights. Rates that year had fallen to $2.50 per mcf, encouraging consumption by around 2,000 households and businesses.[16]

Competition helped secure a portion of what negotiations had not. In 1878, a small amount of natural gas was discovered in the Kansas River. Between 1880 and 1890, city officials awarded five franchises to businessmen promising to deliver the gas, but none acted on the opportunity. A historian of Kansas City later wrote that the franchises were sought "for sale in case some very advantageous situation might

14. King, *History of the Government of Denver,* pp. 152–153.

15. Ibid.

16. Roy Ellis, *A Civic History of Kansas City, Missouri* (Springfield, Mo.: Press of Elkins-Swyers Company, 1930), p. 116; Larsen, *Urban West,* p. 97; Edwin L. Weeks, "Electric Light and Power in Kansas City," in Howard L. Conrad, ed., *Encyclopedia of the History of Missouri,* vol. 3 (New York: Southern History Company, 1901), pp. 362, 365.

perhaps arise." More important, on February 16, 1883, the city government granted a franchise to a Swain and Loomis to erect and operate gasoline street lamps in the outlying areas not served by Kansas City Gas, Light & Coke Company. By December 1884, Swain and Loomis operated 228 lamps. In response, officials of Kansas City Gas, Light & Coke spent more than $27,000 on improvements, including the laying of more than 12 miles of pipe and the installation of 72 new lamp posts, bringing the total to 884.[17]

Up to the mid-1880s, gas companies in Kansas City and Denver were monopolies that thrived on high prices, limited service, and the immense profits thought certain to follow urban growth. Their approach to gas service was essentially an extension of earlier efforts to link modern technology, urban growth, rising property values, and a franchise. At first, city officials in Kansas City and Denver relied on their position as the principal buyers to secure extensions, service improvements, and lower costs. When near monopsony failed to deliver the expected results, city officials turned to competition or the threat of competition with other gas companies. Increasingly, politicians claiming to represent urban growth and a wide-spread interest in gas lighting demanded lower rates and service to residents who were following trolley lines and street improvements to houses and apartments located miles from downtown. Competition with electric lighting imposed the discipline that amateur government could not.

At first glance, electricity offered certain advantages relative to gas. Electric arc lamps provided a brighter illumination than gas. Electricity also flowed through lines that could rapidly be strung above ground rather than through expensive underground pipes that required long periods for installation. Indeed, electricity for lighting purposes was highly amenable to rapid urban growth and to the demands of residents and their political leaders for prompt hookups, maybe even at rates lower than those charged by gas companies. But operators of electric companies also had to adapt to the discipline of rapidly growing cities and to the preferences of local politicians.

Founders of electric companies in Kansas City and Denver had backgrounds similar to those who had launched gas and water companies. During the early 1880s, in fact, several of the investors in the new electric companies were members of the groups who had been

17. Ellis, *Civic History of Kansas City*, p. 116; Kansas City, Missouri, City Engineer, *Report of City Engineer for Calendar Year of 1884* (Kansas City: Press of Ramsey, Millett & Hudson, 1885), p. 84.

active in railroad and real-estate booming, seeking personal wealth and opportunities to boost their city's claim to modernity. Occasionally interests overlapped—in the case of Denver's Walter Cheesman, who helped found the water company and the gas company and then emerged as an investor in the electric company. In Kansas City, to stress a small difference, persons involved in meat packing took an early interest in electricity; in Denver, those accustomed to managing the capital of others were more prominent. Altogether, speculation and commercial striving, more than management, inventiveness, or technology, characterized the interests of members of these groups.[18]

Commercial rivalries and opportunities initiated the development of electric companies. In Kansas City, the impressive lighting and immense crowds at the G. Y. Smith Dry Goods Company had taken place on March 23, 1881, having been advertised in advance as a "grand illuminating display." The next day, local investors, including several persons who operated dry goods and retail businesses nearby, incorporated the Kawsmouth Electric Light Company. (Kaw was a popular name for the Kansas River.) On December 2, 1881, they secured a franchise in perpetuity from the city council and soon began construction of a generating plant.[19]

Commercial rivalries and hoopla also encouraged local investors in Denver to launch an electric company. On April 21, 1880, a lighting demonstration amazed a large number of Denver residents. By early February 1881, City Council promised a franchise to each "company desiring to supply the city with electric lights." Undoubtedly, that promise rested on the perception among council members of the wide popular interest in electric lighting, as well as on a keen appre-

18. PSCC, "Information Concerning the Public Service Company of Colorado" (typescript, c. 1961), p. 6, in Munroe Files; Weeks, "Electric Light and Power in Kansas City," p. 363. Archivist David Boutros at the University of Missouri–Kansas City prepared a list for me of the occupations of investors in the Kawsmouth Electric Light Company. Finally, Forrest McDonald, *Let There Be Light: The Electric Utility Industry in Wisconsin, 1881–1955* (Madison, Wis.: American History Research Center, 1957), p. 50, highlights the decision of political leaders in Milwaukee to encourage competition by "granting franchises to anyone who applied, even when the applicant wanted to serve the same streets and buildings already served by others."

19. [Philip J. Kealy], *Report on the Fair Value of the Property of the Kansas City Electric Light Company and Subsidiary Companies as of February 1, 1914* (Kansas City, Mo.: Smith-Grieves Company, 1914), p. 43, in Kansas City Power & Light, Communication Department, Kansas City, Missouri (cited hereafter as K.C.P&L); advertisement for the G. Y. Smith Dry Goods Company in *Kansas City Evening Star,* March 24, 1881. For the importance of brilliant illumination as part of the rivalry among owners of department stores, see Elaine S. Abelson, *When Ladies Go A-Thieving: Middle Class Shoplifters in the Victorian Department Store* (New York: Oxford University Press, 1989), pp. 86–87.

ciation of the scramble among competitors for franchises that was sure to follow. On February 21, local investors created the Colorado Electric Company and quickly began construction of electrical facilities.[20] In brief, a process of competitive striving between rival groups that had earlier characterized business and politics in Kansas City and Denver had also shaped the contexts within which the new electric firms were established.

Electrical operators still had to turn these businesses into successes in their own right. Determining the location of generators was the first act by which company officials aligned technologies with perceived markets. Similar to the telephone business that was also getting under way, developers of electrical devices assumed that the greatest demand, perhaps all of it, would arise from stores and offices rather than from householders. Only commercial and manufacturing firms would purchase brilliant arc lighting, the reasoning went, because its illumination was too strong and too costly for domestic use. In order to deliver direct current, which flowed only a short distance from crude generators, electrical plants had to be located near likely business customers. In Kansas City, operators constructed their plant at Eighth and Santa Fe in the West Bottoms, an area of cheap rent and yet within electrical reach of railroad stations, packinghouses, and the retail district on the bluffs beyond. The East Side station in Denver, signifying by its name its service area, was constructed at Twenty-first and Wewatta, a district composed of railroad tracks and coal dealers, and equally proximate to likely customers in retailing and manufacturing.[21] From the start, then, operators of electric companies attempted to align technologies with the city's commercial geography.

In Kansas City, electrical operations were under way promptly. During the evening of May 13, a Saturday night, the Kawsmouth plant

20. PSCC, "Information Concerning the Public Service Company of Colorado," p. 6. See also King, *History of the Government of Denver,* p. 154, which offers a different set of dates for the lighting demonstration and for incorporation of the earliest electric companies.

21. Kenneth P. Middleton, "How'd You Like to Pay $30 Monthly for One Electric Light? They Did in 1882," *Kansas City Journal-Post,* February 1, 1925, in Kansas City Public Library. For the perception among early telephone managers that their market would be limited to business uses and household management (such as telephoning plumbers), see Claude Fischer, *America Calling: A Social History of the Telephone to 1940* (Berkeley and Los Angeles: University of California Press, 1992), pp. 65–69. Professor Fischer was kind enough to share several articles and drafts of this book with me. For the limits of direct current transmission up to the early 1890s, see Louis C. Hunter and Lynwood Bryant, *A History of Industrial Power in the United States, 1780–1930, vol. 3: The Transmission of Power* (Cambridge: MIT Press, 1991), p. 244.

supplied current for powerful arc lamps to thirteen stores in the 700 block of Main Street, near the G. Y. Smith Company. Soon Kawsmouth employees, five in all, strung a second circuit looped around the Union Depot south of the plant. By December 1882, Kawsmouth served nearly fifty businesses, including G. Y. Smith.[22]

During the early 1880s, then, groups of investors in Kansas City and Denver launched early electric companies, constructed facilities, and attracted a small number of customers. Apparently, that was about all they could do, given their numerous and more lucrative business dealings and the fact that electrical operations demanded specialized knowledge. According to historian Thomas P. Hughes's lively metaphor, the electrical business was a technological frontier.[23] In order to convert what was still a curiosity and a technical frontier into active and profitable companies, founders turned over day-to-day operations to others who possessed technical training and competence. In turn, those specialists accepted the challenge of adapting electrical operations to fierce competition in rapidly changing cities.

Edwin R. Weeks was the agent of innovation in Kansas City. In 1865, Weeks, a boy of ten, had moved to Kansas City with his parents, a brother, and three sisters. He attended public schools and liked physics in particular, but his first job was delivering newspapers and his second was delivering milk. Next, Weeks withdrew from school and worked as a conductor on the Union Pacific's Kansas City to Denver line. By the mid-1870s, Weeks's parents had died and his brother and sisters were self-sufficient. Weeks enrolled in Phillips Exeter Academy, a preparatory school in Massachusetts, where he became excited about electrical studies. In 1876, a visit to the Philadelphia Centennial Exposition reinforced his interest in electricity. Originally, Weeks had intended to enroll at Harvard College, but in 1882 a shortage of funds and an eye problem that eventually led to blindness brought him back to Kansas City. Simplicity in his personal habits accompanied Weeks's enthusiasm for study and work. As a youth, Weeks was a Seventh-Day Adventist, like his parents, who were also active abolitionists. By middle age, he had declared himself a "practical Christian" and throughout his life remained abstemious in to-

22. Middleton, "How'd You Like To Pay $30 Monthly?"; "History of Electricity in Kansas City, Missouri," *The Tie* 5 (November 1925), 1. (*The Tie* was published for company employees; one set is located at K.C. P&L. I am pleased to acknowledge permission granted by company executives to review *The Tie*.)

23. Thomas P. Hughes, "A Technological Frontier: The Railway," in Bruce Mazlish, ed., *The Railroad and the Space Program: An Exploration in Historical Analogy* (Cambridge: MIT Press, 1965), pp. 53–73.

bacco, coffee, and alcoholic beverages. Weeks's rapid advance from letter carrier to vice president of the largest electric company in Kansas City suggests the cogency of historian Elaine Tyler May's observation that up to the 1880s "the best jobs were usually awarded to those working-class individuals who demonstrated proof of sobriety and piety."[24]

In December 1882, Weeks was appointed superintendent of Kawsmouth's operations. J. D. Cruise, the commercial agent for the Atcheson, Topeka & Santa Fe Railroad and one of the founders of the Kawsmouth Electric Light Company, had recommended to Joseph H. Chick—the company's treasurer, its largest shareholder, and also president of the Bank of Kansas City—that Weeks be given the position. Up to that point, company officials had relied on several shareholders to supervise construction and launch electrical service. Weeks, as superintendent, assumed responsibility for office and plant operations and for bringing additional customers on to company lines.[25]

Investors in the Colorado Electric Company also turned to an outsider for day-to-day management. In 1881, the only person in Denver who comprehended the generation and distribution of electricity as both a technical and a commercial venture was Sidney H. Short, a member of the science faculty at the University of Denver and an inventor. But Short was occupied with development of an electric trolley as well as with the manufacture and marketing of his other inventions, including lighting devices. In the spring of 1881, officials of the Brush Electric Company sent William J. Barker to Denver to direct installation of their equipment. He accepted an offer to remain

24. "Edwin Ruthven Weeks," *Encyclopedia of the History of Kansas City,* 6:427–429; *Kansas City Times,* August 18, 1938, in Kansas City Public Library; Elaine Tyler May, *Great Expectations: Marriage and Divorce in Post-Victorian America* (Chicago: University of Chicago Press, 1980), p. 19.

25. Weeks, "Electric Light and Power in Kansas City," pp. 362–365; "Edwin Ruthven Weeks," p. 427; *Kansas City Journal-Post,* February 1, 1925; *Kansas City Times,* August 18, 1938; [Kealy], *Report on the Fair Value,* pp. 25–26, 103. Because expertise in electrical matters was scarce in areas like Kansas City, Weeks turned to electrical inventors and manufacturers such as Thomas A. Edison and Elihu Thomson for advice on solving critical technical and financial problems. The only other person in Kansas City who had a sophisticated understanding of electrical operations was Gerald W. Hart, a graduate of the Sheffield School at Yale College and salesperson for the American Electrical Company. Founders of the Kawsmouth Company had purchased a license at a cost of $4,000 for the right to use Thomson-Houston equipment. For Gerald W. Hart, see *New Britain Herald,* April 25, May 16, 1882, March 10, 1931, in New Britain Public Library, the notes for which W. Bernard Carlson was kind enough to share with me. For the concept of critical problems in the development of electrical systems, see Hughes, *Networks of Power,* pp. 14–15.

as engineer, later assuming positions of broader responsibility. Barker in Denver, and Weeks in Kansas City, interpreted technical, organizational, political, and urban change for officers and investors.[26]

At first, Weeks and Barker enjoyed no more than modest success in bringing electric lighting to their cities. In Denver, the gas company had an exclusive contract to light the city's streets. Because of the brilliance of arc lighting, the few private installations that could be done tended to be limited to dry goods stores and saloons, where managers and owners wanted a novelty effect for competitive purposes. By late 1881, no more than 200 arc lamps hung indoors. Officers of the Colorado Electric Company settled on two strategies, each probably known to Barker through his former association with the manufacturer. First, Colorado Electric erected seven towers, each 150 feet high, and hung six 3,000-candle-power arc lights on top. Everyday, workmen ascended the towers in small cages to service the lamps. Nicknamed the "lighthouses of the plains," in reality the towers offered an impressive demonstration of the brilliant illumination that arc lighting could provide, and did so in the fast-growing areas outside downtown and outside the gas company's exclusive district. Second, after 1882, Colorado Electric provided lighting on a portable basis for parties and other evening gatherings. Horse-drawn wagons delivered large batteries, wires, and Swan incandescent bulbs, removing them in the morning and subsequently recharging the batteries.[27]

High prices and the gas company's exclusive franchise for street lighting blocked installation of arc lamps for street lighting in Kansas City. As Barker had done in Denver, Weeks determined to solicit contracts from merchants for whom a precise calculation of gas and electric costs apparently mattered less than the intangible value of featuring lighting by electricity. Costs were high. Kawsmouth charged $15 a month for each lamp in service from dusk until 9:00 P.M.; all night lighting cost $30 per lamp. In 1884, Weeks created a rate of

26. "Light of the Future," *Denver Republican,* April 1, 1881, SHSC; Joseph E. Smith, "Personal Recollections of Early Denver," *Colorado Magazine* 20 (March 1943), 66; "Sidney Howe Short," *Dictionary of American Biography,* vol. 9, p. 128; Harold C. Passer, *The Electrical Manufacturers, 1875–1900: A Study in Competition, Entrepreneurship, Technical Change, and Economic Growth* (Cambridge: Harvard University Press, 1953), pp. 14–19; PSCC, *Public Service Company of Colorado: Its Past and Its Present* (n.p., c. 1976), p. 6; "Who's Who in the Doherty Organization," c. 1920, Munroe Files. W. Bernard Carlson called Sidney H. Short to my attention.

27. "The Plant of the Colorado Electric Company at Denver, Colo.," *Electrical World,* September 1, 1888, p. 103 (W. Bernard Carlson located this item as part of his own research and shared it with me); PSCC, *Providing Energy for More Than a Century,* p. 7; Barth, *Instant Cities,* p. 223; Passer, *Electrical Manufacturers,* pp. 19–21.

$18 for service until 10:00 P.M., and another rate of $22 for midnight lighting—all still on a monthly and per lamp basis. By way of contrast, gas, for outdoor lighting, cost $30 per lamp for an entire year, though gas lighting was far less luminescent. In July 1885, Weeks had secured contracts for about 250 arc lamps that were operating on several circuits in the central business district, including Grand Avenue from Third to Fifteenth street.[28]

Rapid expansion of electric street-lighting began in 1885, after an explosion at the plant of the Kansas City Gas, Light & Coke Company halted gas-making operations. Stores and streets were suddenly darkened, and residents had to purchase kerosene lamps to secure short-term relief. As recently as April 1885, Weeks and senior officers in the firm had failed to receive council's approval to place 100 arc lamps in the central business district. Gas officials had warned City Council that awarding a contract to the electric company would force an increase in rates for each of the 572 gas lamps located outside of downtown. Within several weeks of the explosion, however, the electric company installed a circuit and twenty-five arc lamps, the first outdoor electric illumination in the city. By the end of January, 400 arc lamps were operating in stores and on the streets of Kansas City, one of the largest number in the nation.[29]

In 1885, electric street-lighting also was introduced in Denver. The origins of this change were less dramatic than an explosion at the gas works, but more sweeping in their results. In 1885, the contract between the city and the gas company expired. In one swoop, members of City Council awarded the Colorado Electric Company a contract to light the entire city. Politicians were taking back what they had given. The electric company replaced each gas light with a 20-candlepower Edison incandescent lamp. By mid-1888, 900 incandescent bulbs and seven arc towers, the "lighthouses of the plains," lit Denver's outdoors, while about 450 Brush arc lamps and Westing-

28. Robert A. Olson, *Kansas City Power & Light: The First Ninety Years* (New York: Newcomen Society in North America, 1972), pp. 9–10; *Kansas City Journal-Post,* February 1, 1925; "Arc Plants," in Notebooks, Elihu Thomson Papers, Library of the American Philosophical Society, Philadelphia (which W. Bernard Carlson identified and copied for me); "History of Electricity in Kansas City, Missouri," p. 1; K.C. P&L, unpublished history of Kansas City Power & Light (manuscript in the author's possession), c. 1982, p. 8; [Kealy], *Report on the Fair Value,* p. 26. The smallest arc lamp provided 1,200 candlepower of illumination.

29. Ellis, *Civic History of Kansas City,* p. 117; *Electrical World* 5 (June 6, 1885), 225; Weeks, "Electric Light and Power in Kansas City," pp. 363–364; "Arc Plants"; *Kansas City Star and Times,* May 12, 1925, in Robert M. Snyder Jr. Collection, University of Missouri at Kansas City (cited hereafter as Snyder Collection).

house incandescent bulbs were illuminating interior spaces. According to a report in *The Electrical World,* the Colorado Electric Company was already "lighting districts so remote that it will be years before the gas company can cover this territory with their mains."[30]

Until the 1880s, leaders in the gas industry had assumed the permanence and preeminence of manufactured gas for lighting. In most of the larger cities of North America, including Kansas City and Denver, gas had the advantage of being a familiar technology, and preferable to candles and kerosene. By the late 1880s, gas had been dislodged from street lighting in Denver and had to share a place with electricity in Kansas City. The growing popularity of incandescent lighting for the illumination of interior spaces also threatened gas; worse yet for the gas interests, rapid stringing of electric lines to the periphery of each city portended elimination of future markets. During the 1890s, however, lower rates for gas and several innovations in gas manufacture fostered intense competition between gas and electrical operators, leading to a temporary restoration of the gas companies to superior positions in local markets.[31] In turn, these battles of systems, technologies, and prices exhausted treasuries as well as political resources.

Up to the 1870s, gas companies in virtually every city had manufactured gas by heating coal. As an illuminant, however, coal gas smelled, created soot, and gave a dull light in comparison with kerosene and incandescent bulbs. Beginning in the 1870s, gas operators in many cities adopted a new method of making gas. In brief, gas manufacturers began to produce water gas, the result of forcing steam across the coke beds and then adding oil vapors. Water gas burned much brighter than coal gas.

30. "Plant of the Colorado Electric Company," pp. 103–104.

31. Passer, *Electrical Manufacturers,* p. 174. Harold L. Platt, *The Electric City: Energy and the Growth of the Chicago Area, 1880–1930* (Chicago: University of Chicago Press, 1991), pp. 23–24, observes that historians regularly characterize this early period in electrical operations as one of a "battle of the systems." He also points out, correctly I think, that the battle was not simply between operators of alternating and direct current plants but also between electrical operators, gas operators, and persons operating small electrical plants located in factories and office buildings. During the course of this battle of the systems in Kansas City and Denver, owners of hotels, office buildings, and factories installed electric generators, and members of city councils granted franchises to several firms operating gas and electric systems. In order to maintain a focus on technology, public policy, and urban growth, however, I have identified the gas and electric firms that launched service in those two cities and that survived up to 1900. For accounts of franchises and new gas and electric firms in Kansas City and Denver

The second innovation was the Welsbach lamp. Named after its inventor, Carl Auer von Welsbach, this lamp used fabric embedded with rare earths. The fabric, set in an ignited gas lamp, became incandescent, creating an illumination six times that of the gas jet. In 1890, the United Gas Improvement Company of Philadelphia purchased the American patent rights to the Welsbach lamp from its German inventor and began manufacturing them. Gas companies also sought new markets for their product. As early as 1884, several gas companies in New York were cooperating in marketing gas stoves as a replacement for the coal range. Between 1878 and 1888, moreover, gas prices in a number of cities dropped about 50 percent. Competition with electric firms required gas operators "to take positive steps to survive." In Kansas City and Denver, those positive steps varied according to perceptions of public policy, patents, and urban growth.[32]

Operators of the Denver Gas Company chose a defensive strategy based on the acquisition of patents. In September 1889, gas officials had opened a new manufacturing plant consisting of twenty-two benches, each with a capacity of 50,000 cubic feet a day. But it was an old-fashioned coal-gas plant. In 1891, leaders of the Denver Gas Company purchased the United Gas Improvement Association and the People's Gas Light Company. In December 1891, the Denver Consolidated Gas Company emerged as the operating and legal entity for gas supply in Denver. Consolidation was not aimed at removing competition in the gas business; none existed. Instead, gas company officials were actually purchasing patent and franchise rights. In particular, the United Gas Improvement Association held rights to water-gas equipment, and of course to the Welsbach lamp. By 1893, installation of two water-gas machines with a capacity of 600,000 cubic feet a day and a storage tank with space for another 100,000 cubic feet provided the technical prerequisites for low-cost and around-the-clock operations against electricity.[33]

The political economy of gas took a different turn in Kansas City. Originally, public officials relied on their position as the largest consumer of gas to secure reduced rates for everyone. Since 1867, the Kansas City Gas, Light & Coke Company had been doing business in the city under a charter provided by the state. By 1894, city govern-

---

during this period, one must continue to rely on the reports of contemporaries. See, for example, King, *History of the Government of Denver;* [Kealy], *Report on the Fair Value.*

32. Passer, *Electrical Manufacturers,* pp. 195–198.

33. "History of Public Service Company Gas Department," 1958 (typescript), folder entitled "Gas Department History, 1958," in PSCC Records; PSCC, *Public Service Company of Colorado,* 8; Frueauff, "History of the Company Gas Properties."

ment had persuaded company officials to reduce rates from $4.50 a thousand to $1.60.[34]

Following the depression of 1893, aldermen and mayor alike in Kansas City made competition the preferred strategy in the gas business. In 1895, the company's charter would expire and politicians were eager to secure relief. Rates charged the middle classes locating in the fast-growing districts south and east of downtown were at the heart of the matter. That gas facilities in Kansas City were owned by the United Gas Improvement Company (UGI) of Philadelphia made a skirmish with the beast more attractive. By October 1894, the president of City Council was proclaiming the advantages of competition. Now, leaders in city government liked the prospect of soliciting competitive bids on the franchise as a device both for lowering rates and for directing a portion of the company's revenues to city government. Nonetheless, the conviction among officials was that "the day of franchise giving has passed."[35]

Robert M. Snyder emerged as the operator of Kansas City's second gas company. Snyder had been raised in Louisville, Kentucky, held positions as a clerk and merchandise broker, and moved to Kansas City in 1880. In Kansas City, he entered the wholesale grocery business. By 1884, Snyder was plunging into the real-estate craze, but balancing his holdings with investments in cattle and grazing lands. One of his biggest deals was the purchase of an option on the land where the new city hall was to be constructed, which turned over at a price of $450,000. Snyder also sold securities, which led in 1890 to the formation of the Mechanics Savings Bank, with himself as president. In short, Snyder was part of the tradition of business leaders

34. Ellis, *Civic History of Kansas City*, pp. 116–117.

35. Ibid., p. 117; *Kansas City Star*, October 22, 1894; See also David Thelen, *The New Citizenship: Origins of Progressivism in Wisconsin, 1885–1900* (Columbia: University of Missouri Press, 1972), pp. 225–229, for an account of the development of a consumers' movement in Wisconsin following the depression of 1893. For the appearance in larger cities of a distinctive and self-conscious middle class during the nineteenth century, see Stuart M. Blumin, *The Emergence of the Middle Class: Social Experience in the American City, 1760–1900* (New York: Cambridge University Press, 1989). With their promise of lighting the night, gas and electric service were the exception to the rule before 1900 of modest tax levies and modest expenditures by urban politicians. See, for example, Jon C. Teaford, *The Unheralded Triumph: City Government in America, 1870–1900* (Baltimore: Johns Hopkins University Press, 1984), p. 218. For the origins and programs of limited government, see Terrence J. McDonald, *The Parameters of Urban Fiscal Policy: Socioeconomic Change and Political Culture in San Francisco, 1860–1906* (Berkeley and Los Angeles: University of California Press, 1986), pp. 203–238.

who found gas and electric operations a worthy extension of booming and boosting in commerce, banking, real estate, and politics.[36]

In January 1895, Snyder and the city reached agreement regarding rates that his new company would charge and the size of the area it would service. Gas would be available for $1 per mcf. In addition to a low price, the company was required to return 2 percent of gross revenues to the city and to provide gas for street lamps at $18 a year. No one did better under the terms of this agreement than residents of sparsely settled sections of the city. Snyder was obligated to provide service to as few as six customers located within 400 feet of one another.

Leaders of the Kansas City Gas, Light & Coke Company—the "old company," as it came to be called—had offered gas for $1.40. A rate any lower, they complained, amounted to confiscation. But with Snyder's contract to deliver gas at $1 in hand, city officials quickly turned aside the old company's offer. By August 1895, leaders of the old company had agreed to identical terms plus payment of the city's legal fees deriving from the two lawsuits that executives of the old company had brought. Even more, leaders of the old company agreed to make refunds of charges in excess of $1 collected since March 1, as a form of penance.[37] All who purchased gas, including the taxpayers who financed the city's bill, would presumably benefit from these rates and service requirements, especially householders in the sparsely settled subdivisions located south and east of downtown. Consumer and ecological politics were to have their first day, and they did.

The price war started before Snyder got his plant into operation. By March 1896, the old company, which earlier had claimed it could not sell gas for less than $1.40, was now offering it for 50 cents. In addition, the old company would supply free of charge a stove and public lectures by a Miss Andrews on using it. Free plumbing and free installation of lamps and stoves became the working rule in the mayhem of attracting customers. Gas at 50 cents was below cost, but Snyder had raised and spent approximately $800,000 based on signed contracts, some 16,000 in all, to purchase gas for $1. Sooner or later, Snyder would have to ask customers to make good on those contracts.

On March 28, the Missouri Gas Company, Snyder's new firm, began to pump gas through 65 miles of pipe, much of it lying south of

36. "Robert M. Snyder," *Encyclopedia of the History of Missouri,* 6:13.

37. Kansas City, Missouri, Public Utilities Commission, *Laws, Ordinances, and Permits* (1912), pp. 121–141. The Hauk-McGurren Gas Company of Chicago also offered to provide gas for $1. See *Kansas City Star,* October 15, 1894.

downtown. Civic and religious leaders joined with Snyder and his executives in optimistic pronouncements that contracts for gas at $1 would be honored despite the old company's offer of gas at 50 cents. "The [U.G.I.] trust is presuming . . . upon the . . . cupidity and stupidity of the people," contended attorney and civic leader J. M. Lowe, but "there is no danger of this." The Rev. S. M. Neel, pastor of the Central Presbyterian Church, reported that he planned to "patronize the new gas company at a dollar in the face of the old company's price of 50 cents, because I think it would be morally wrong to assist the old company in their attempt to injure the company who made dollar gas possible." Snyder told a crowd on opening day that the gas works was constructed on "faith in these pledges . . . , and I stand here to say these pledges will be kept."[38]

The price war ran its course in ten months. In February 1897, leaders of the two companies petitioned council for authority to merge operations. The mayor vetoed the ordinance on the grounds that competition benefited consumers. Council overturned him, citing the inevitability of consolidation, the reasonableness of the charge of $1, safety for capital, and Kansas City's small population and large territory. Still another council member suggested holding a "mass meeting . . . [to] declare R. M. Snyder the grandest philanthropist that Kansas City has ever had." Until 1906, the old company would be able to sell gas for $1 in Kansas City.[39]

Cheap gas had contributed to the destabilization of the electric lighting business in Kansas City and Denver. Yet price wars were only one of several factors destabilizing these companies. Insistence by city governments on service to residents relocating to sparsely settled districts along the urban periphery added to the burdens of small and poorly capitalized firms. By the mid-1890s, as an example, Weeks had extended electric lines a distance of 5 miles into the thinly settled suburbs of Kansas City. Beginning around 1897, the convergence of new technologies, public policy, and rapid out-migration only led to additional mergers, another round of price wars, and bankruptcies.[40]

38. *Kansas City Journal,* April 5, 1896; and newspaper clipping c. February 1897, both in Snyder Collection; *Kansas City Star,* March 29, 1896; *Kansas City World,* March 29, 1896; [Kealy], *Report on the Fair Value,* p. 29.

39. Clipping c. February 1897; Kansas City, Missouri, Public Utilities Commission, *Laws, Ordinances, and Permits,* pp. 141–143; Carrie Westlake Whitney, *Kansas City, Missouri: Its History and Its People* (Chicago: S. J. Clarke Publishing Company, 1908), p. 275; Robert M. Snyder Jr.'s chart of gas prices in Kansas City, 1890–1931 (in Snyder Collection).

40. "Electric Light and Power in Kansas City," p. 365.

Devolution of the electrical business in Kansas City began in 1896 and continued to the end of the decade. Several interlocked firms under the direction of Edwin Weeks controlled most of the electrical capacity in Kansas City. Late in 1898, however, with only 23,000 incandescent bulbs installed in the entire city, that capacity remained underutilized. Without doubt, cheap gas at 22 candlepower appeared more attractive than costly and mysterious electric lighting.[41]

The losses suffered by operators of the Kansas City Electric Light Company (formerly Kawsmouth) highlight the dismal state of the electrical business in Kansas City. Between 1881 and 1896, Weeks had spent more than $669,000 on plant and equipment. Bonded indebtedness amounted to $300,000, half of which had been issued in 1894 as part of the purchase of the American Electric Light Company, an electric operation started by executives of the old gas company. Thereafter, interest paid on bonds (and other debts) consumed about 45 percent of net income. Beginning in 1896, directors passed on dividend payments. In 1897, moreover, the bank in which company funds were deposited failed. More revealing of management's perception of the company's future was the fact that combined investments for 1898 and 1899 amounted to less than $4,000.[42]

Although documentation is not explicit on this matter, it appears that by late 1899 Weeks set two courses of action before the directors. On the one hand, they could increase the scope of operations, including construction of a universal generating plant and a series of substations. Universal systems held out the promise of increased efficiency in serving diverse and decentralizing cities. Such a universal system would replace the conglomeration of inefficient equipment at work in Kansas City (and in Denver). In 1893, employees of the Westinghouse Electric & Manufacturing Company had demonstrated this system at the World's Columbian Exposition in Chicago. In 1896, Westinghouse engineers installed the fabulous system of generators and lights that

41. *Kansas City Star,* October 15, 1898, in clipping files, Kansas City Public Library. One must rely on different estimates of the number of electric lights in service. The report of the city's Board of Public Works for 1896 listed 209 arc lights and no incandescent bulbs in use to light city streets, suggesting a limited demand from public sources for electric lighting during a period of declining rates for gas lights. By 1900, according to Weeks, about 100,000 incandescent bulbs and 3,000 arc lamps were burning in Kansas City. But that figure, as he pointed out, included lamps in private use that were powered by independent electric plants located in factories, hotels, and office buildings. Compare "Electric Light and Power in Kansas City," p. 366, with *Eighth Annual Report of the Board of Public Works of Kansas City, Missouri, for the Fiscal year 1896* (Kansas City, Mo.: H. S. Millett Publishing Company, 1897), p. 11.

42. [Kealy], *Report on the Fair Value,* pp. 29, 33, 103, 105.

created so much excitement at Niagara Falls. But construction of a universal system required immense capital. On the other hand, directors could vote to sell their holdings to another firm that had access to the necessary credit and funds. Joseph Chick, one of the founders, wanted to liquidate his holdings—which apparently decided the matter. Directors signed their stock certificates in blank and authorized Weeks to negotiate a sale on their behalf. In January 1900, J. Ogden Armour purchased the holdings of the Weeks group and commenced the task of achieving an improved fit among public policy, a decentralizing population, and his electrical technologies.[43] In the long run, Armour's luck was no better.

The fragmentary record of the gas and electric companies in Denver highlights a similar pattern of competition, declining prices, and consolidation in the hands of outsiders. During the 1890s, competition between the city's gas and electric companies for a limited amount of lighting business continued without relief. Rather than continuing to expend scare resources, however, leaders at both firms initiated a strategy of opening new markets. Officials of the gas company, for example, sought to develop the heating market, even in the face of mountains of cheap coal along the Front Range. Executives of the electric company attempted to develop the power market. But manufacturers were reluctant to abandon investments in steam engines and belts—both of which had the added advantage of familiarity—for the expenses and uncertainties of electricity and direct drive. In the mid-1890s, such manufacturers as General Electric and Westinghouse were only beginning to conduct the research and development that were prerequisite to production of inexpensive and reliable electric motors. Finally, the national depression that began in 1893 proved especially severe in Colorado, adding to the economic and technological inertia.[44]

43. "Edwin Ruthven Weeks," p. 427; "Electric Light and Power in Kansas City," p. 366; Olson, *Kansas City Power & Light,* p. 12. For the cultural meaning of electric lighting at the World's Columbian Exposition, see Platt, *Electric City,* pp. 59–65, 81; and for the meaning of electric lighting at Niagara Falls, see Nye, *Electrifying America,* pp. 58–60. Finally, Hughes, *Networks of Power,* pp. 122–127, 137–139, remains the standard account of the relationship between technical innovations, corporate mergers, and a growing demand for electric service.

44. "The Denver Consolidated Electric Company," c. 1892, in Irving Hale Papers, Box 9, FF 555, Denver Public Library (cited hereafter as Hale Papers). Passer, *Electrical Manufacturers,* pp. 296–318, and Hughes, *Networks of Power,* pp. 140–174, assess research among the manufacturers of electrical equipment. Hughes also describes the period of the 1890s as one of technological momentum, which was no doubt the case for manufacturers and for electrical developments as a whole. In Denver and Kansas

Restoration of economic activity around 1896 only attracted the attention of potential competitors in the production of gas and electricity for a modest-sized market. Late in 1896, J. E. Rhodes announced a scheme to build a hydroelectric plant north of Denver, delivering electricity for lighting, trolley service, and power. Rhodes had developed electric service in Ogden, Utah, serving both the trolley and street lighting. Irving Hale, director of General Electric's office in Denver and formerly the engineer for the electric trolley in the city, had gathered the engineering and economic data on which the proposal was based. As in other cities in which universal systems were being planned and developed, developers promised lower prices. "We come in competition with cheap coal and gas that produce cheap power," Rhodes announced, and "the competition thus established . . . will enable those in the manufacturing business to secure power cheaper than any point west of the Mississippi River."[45] The project never reached fruition, but the fact that it had matured to the point of seeking a franchise from the city highlights the shallowness of the political and technological links of both the electric and gas operations up to that point. Public policy broke those links.

Consumer politics was entitled to its day in Denver. By the mid-1890s, the urge to regulate utility rates had reached the Rocky Mountain region. The depression of 1893, as elsewhere, had encouraged consumers and politicians to seek control of rates through the political arena. Monopoly, it was argued, had made regulation impossible, and only municipal ownership could force rates down to acceptable levels. Many contended that municipal ownership would allow the city to reduce its lighting costs up to 80 percent. Members of the state senate, gripped with reformist fever, authorized city and town governments to construct gas, electric, and water plants, with voter approval, or to purchase those plants following a number of years of operation. By late 1899, even the mayor of Denver had embraced the idea of securing lower rates through municipal ownership, indicating just

---

City, however, utility operators lacked research skills and resources. Members of the original groups of owners, along with Weeks and Barker, operated plants on the basis of practical experience and with the assistance of practical mechanics, rather than with the tools and logic of mechanical and electrical engineers.

45. Irving Hale to Denver Consolidated Tramway Company, and J. E. Rhodes, February 7, 1896, FF 517; "Estimated Cost of Plant Having a Capacity of 1000 Arc Lights (800 Street, 200 Commercial) and 150 miles line," c. 1896, FF 555; and "Prospectus of the Denver Power and Irrigation Company," December 29, 1896, and clipping, c. January 1897, both in FF 528—all in Hale Papers.

how widespread was the passion animating consumer politics in the electric lighting field.[46]

Early in 1900, city government joined in the general enthusiasm for securing cheaper prices and dismantling giant corporations. On March 30, 1900, members of Denver's council voted a franchise to the LaCombe Electric Company. Current for arc lamps would fall to $90 a year, a reduction of $30 from the current price. Household rates, unregulated but costing 15 cents per kilowatt hour (kwh) would drop to 10 cents. Still more, the organizer of the company, Charles F. LaCombe, agreed in advance to sell the arc plant to the city at a declining price during each of the next ten years. The commercial lighting plant would also be available for purchase in ten, fifteen, or as long as twenty years. Payment of a portion of gross revenues on the lighting plant, *de rigueur* in the new political climate nationwide, was fixed at 3 percent.[47] In short, a franchise granted years earlier under different circumstances was to be set aside in the frenzy to purchase electric lighting at a lower price from a more vulnerable company.

The politics of municipal ownership comprised only the final step in the erosion of small and independent gas and electric operations in Denver. Beginning in January 1899, the gas and electric companies in Denver rejoined the movement toward merger of local utilities and consolidation into national corporations. Emerson McMillin, a banker located in New York City and the "light king of this country" according to a Denver newspaper, purchased shares of both the gas and electric companies and merged them into the offices of the gas company. In the local battles of rates, illuminating power, and service to the periphery of the city, the gas company was the survivor and apparently the victor. McMillin promised an infusion of cash, greater efficiency, and higher profits to the remaining shareholders, and lower prices for the rate payers, including free hookups. On June 1, 1899, the board of directors of the new Denver Gas & Electric Company, which included officers of the last electric firm, lowered gas rates to $1.35 a thousand, a reduction of 15 cents.[48]

46. *Evening Post,* January 12, 1895 (microfilm); *Denver Republican,* March 5, 1899, and March 28, 1899; *Denver Times,* February 3, 1899, and December 15, 1899, all at the Denver Public Library, the last four in the clipping files. Beginning in 1895, Denver paid nearly $200,000 a year for street lighting. See King, *History of the Government of Denver,* p. 207.

47. King, *History of the Government of Denver,* pp. 208–209.

48. *Denver Times,* February 3, 1899, June 4, 1899 (both on microfilm), June 7, 1899, and June 8, 1899 (both in clipping files), all at the Denver Public Library.

The city's award of a franchise to Charles LaCombe initiated another rate war. On March 16, 1900, members of the board of directors of the new Denver Gas & Electric Company held a special meeting. Their object, according to the notes of the company secretary, "was to consider and discuss the pending franchises and the position of the company regarding them." Directors dropped the price for arc lighting to $75 a year, a rate of $15 below LaCombe's. Fierce competition continued during the next two years, driving rates as low as 2.5 cents per kwh for householders burning incandescent bulbs.[49]

The net results of this excursion in competitive fervor extended beyond financial losses. Founders of gas and electric companies in Kansas City and Denver were accustomed to risky investments. They had selected banking, retailing, and real estate in remote cities possessing small populations. Naturally, investments in gas and electric companies appeared to be another opportunity to get a jump on local competitors and to make a familiar set of connections between urban growth, a public franchise, the glamour of technological modernity, and a large profit.

One irony was that founders of these urban gas and electric companies secured what they had been seeking but then were overwhelmed by it. By the mid-1890s, investors such as Joseph Chick and managers such as Edwin Weeks and William Barker were learning that urban growth and changes in political outlook turned into demands to provide low-priced service, often to householders living far from downtown. From the point of view of members of the founding generation, creating a new set of links between costly universal electric systems and an increasingly unreliable set of politicians appeared a risk not worth taking. By 1900, members of these groups had determined that it was preferable to manage capital rather than technologies and public policy.

By 1900, directors possessing immense resources had taken charge of gas and electric firms in Kansas City and Denver. New managers quickly increased the size and scope of operations. In Kansas City, operators of the gas company brought cheap and high-quality natural gas to the city and agreed to the demands of politicians that they make connections to houses located in far-flung districts. In Denver, Emerson McMillin sent Henry L. Doherty to direct the combined gas

49. Minutes of the director's meeting, March 16, 1900, in Record of the Meetings of the Denver Gas & Electric Company, 1899–April 30, 1909, Doc. 35113X, p. 104, PSCC (cited hereafter as DG&E Meetings).

and electric company and to fight the rate war against the LaCombe company. In both cities, the emphasis was on adapting gas and electric operations to demands for cheap rates and distant service. In the long run, only Henry Doherty had the skill and the luck to link politics, technologies, and urban change.

CHAPTER TWO

The Blunt Discipline of Public Policy, 1900–1920

> In sum, public utilities, both national and local, changed in response to growing markets and changing technology in the last half of the nineteenth century. But the nature of the urban market differentiated the experience of local utilities by involving them more closely in politics and policy.
>
> —Historian Charles W. Cheape, 1980

BEGINNING AROUND 1900, political leaders in many cities expected officials of gas and electric companies to provide inexpensive service to residents, especially to residents of sparsely settled districts along the city's periphery. Political leaders created part of the framework for mass production and consumption of gas and electricity. In Denver, Henry L. Doherty fixed a program of rates and mass marketing that satisfied politicians, undercut competitors, and eventually won a new franchise. In Kansas City, however, new groups of managers could not fit gas and electric operations to the politics of growth until after World War I.[1] Monopoly status, modern technologies, and the

1. The political economy of gas, electric, and trolley corporations in American cities between 1900 and 1920 is an area of study where findings remain incomplete and contradictory. One question dividing historians has been whether political leaders or utility company executives directed the process of securing and implementing regulation. In brief, where was authority located? Thomas P. Hughes, *Networks of Power: Electrification in Western Society, 1880–1930* (Baltimore: Johns Hopkins University Press, 1983), pp. 18–19, accords to executives like Thomas A. Edison and Samuel Insull the ability to coordinate technical, legal, economic, and political innovations. By contrast, Christopher Armstrong and H. V. Nelles, in *Monopoly's Moment: The Organization and Regulation of Canadian Utilities, 1830–1930* (Philadelphia: Temple University Press,

resources of great corporations could not ensure the survival of gas and electric firms in urban politics.

Only the outlines of Henry Doherty's youth and early adulthood are known. In fact, much of what we know about Doherty comes from his own recollections, which were inevitably colored by time and by the uses to which he put the past in correspondence with colleagues, superiors, and subordinates. Doherty's employees prepared information about his early career, but that material is simply adulatory. Even so, it is possible to piece together a sketch of the skills, attitudes, and accomplishments that he brought with him to Denver.

In 1870, Henry L. Doherty was born in Columbus, Ohio. He remained in school until age twelve. "I could not get along in school," he wrote years later, "and was under expulsion a large part of the time I was supposed to be in school." Still age twelve, Doherty went to work at the local gas company. Thereafter, in his own self-flattering phrase, "I have been entirely self-educated." By 1896, Doherty had so impressed management with his talents that he was asked to take charge of one of their companies, a declining gas and electric operation in Madison, Wisconsin. According to the legend that came to surround Doherty (as well as successful executives like him), he accepted that position on the spot, catching "the next train out of Columbus."[2]

During the next few years, Doherty began to articulate some general ideas regarding rates and promotional activities that took account

---

1986), argue that the precise authority of politicians and executives in regulatory matters in Canada was contingent on certain circumstances—for example, strikes, the French-English rivalry, and political rivalries within Toronto, Montreal, and Vancouver. Analyses of the literature of technology, cities, and public policy are in Robin L. Einhorn, *Property Rules: Political Economy in Chicago, 1833–1872* (Chicago: University of Chicago Press, 1991), pp. 12–27; Josef W. Konvitz, Mark H. Rose, and Joel A. Tarr, "Technology and the City," *Technology and Culture* 31 (April 1990), 284–294; and Mark H. Rose, "Machine Politics: The Historiography of Technology and Public Policy," *Public Historian* 10 (Spring 1988), 27–47.

2. Henry L. Doherty to Dr. S. W. Stratton, November 27, 1930, MIT Archives and Special Collections, Collection AC 13, Box 7, Folder 182, Cambridge, Mass.; Forrest McDonald, *Let There Be Light: The Electric Utility Industry in Wisconsin, 1881–1955* (Madison, Wis.: American History Research Center, 1957), p. 76; Glenn Marston, comp., *Principles and Ideas for Doherty Men: Papers, Addresses, and Letters, by Henry L. Doherty,* vol. 1 (New York: Henry L. Doherty & Company, 1923), p. 20. For an analysis of the myth of the self-made executive, see R. Richard Wohl, "The 'Rags to Riches' Story: An Episode in Secular Idealism," in Reinhard Bendix and Seymour Martin Lipset, eds., *Class, Status, and Power: A Reader in Social Stratification* (Glencoe, Ill.: The Free Press, 1953), pp. 388–395.

of politics, cities, and profits. As early as 1894, he had prepared a paper focusing on the principles of rate-making. He contended that each customer ought to pay a monthly bill made up of three parts rather than the usual flat rate or a metered rate. First, customers had to pay for the amount of gas and electricity they actually consumed; second, he thought they ought to pay a proportionate share of the costs of the equipment that stood in waiting should they turn on lights or stoves; and third, he figured that a hookup fee charged monthly was also in order. By 1900, Doherty had prepared several tables relating the details of expenses to potential income under his three-part rate, suggesting an analytical style similar to that of Thomas A. Edison in an earlier period and to his contemporary Samuel Insull in Chicago. For Doherty, rate-making was part of a strategy for dealing with the tough competition posed by coal and kerosene dealers and the uncertainties posed by proponents of lower rates, including advocates of municipal ownership. According to Doherty, when he spoke to electric plant operators at their national convention in 1900, his ideas would "develop . . . business, repress agitation for municipal ownership . . . , and . . . prove a panacea for most of the ills to which the average central station is heir."[3] Up to 1900, however, those ideas about rates as part of a broader urban strategy had not been implemented in the political context of a large market.

On October 18, 1900, Doherty became the acting president and treasurer of the Denver Gas & Electric Company. Good luck as well as good works had brought him that far. In June 1899, Emerson McMillin (the "light king," as described in the previous chapter) had sent George T. Thompson to Denver with a charge to serve as president of the recently consolidated gas and electric company. Thompson had been associated with gas and electric operations for about fourteen years, including ten years in St. Louis and the last three as one of McMillin's senior executives. But Thompson died on October 1, 1900, and McMillin asked Doherty to replace him.[4]

Chance and luck are a part of the historical process. One must also suppose that Doherty's obvious talents would have found an audience in the near future. But Thompson's appointment, and then Doherty's, were also expressions of a more general pattern that was beginning

3. Marston, *Principles and Ideas,* pp. 10, 20; Henry L. Doherty, "Equitable, Uniform, and Competitive Rates," *Proceedings of the National Electric Light Association, 23rd Annual Meeting* (New York, 1900), pp. 291–321. For Edison and Insull, see Hughes, *Networks of Power,* pp. 18–46, 201–226.

4. *Denver Times,* June 8, 1899, clipping files, Denver Public Library; DG&E Meetings, pp. 118–119, 121.

Fig. 1. Henry L. Doherty, President, Denver Gas & Electric Company, no date. Doherty was one of the first utility executives to align rates, distribution systems, and marketing with the demands of political and business leaders for uniform rates on a citywide basis. (Courtesy, Public Service Company of Colorado)

to emerge in gas and electric operations in a number of cities by about 1900. Those with experience in urban technology and politics were replacing electrical enthusiasts and real-estate boosters in the day-to-day management of electric and gas firms.

Doherty's subsequent success in Denver stemmed from his keen sense of the political geography of rate-making, from his technological and organizational virtuosity, and from the intense marketing campaigns that he directed. Indeed, Doherty created an organization that was capable of diffusing technological knowledge and adapting to urban change. Adjusting rates, then, was only a first step in boosting gas and electric utilization among the well-to-do who were moving to the city's edge. They comprised the most active group seeking tough regulation of rates and even municipal ownership.

In November 1900, Doherty sketched a general scheme for implementing rate reductions and encouraging increased consumption. Earlier in the year, members of the company's board of directors had ordered a small reduction in the rates charged householders and the city for electric lights. Now, Doherty outlined his three-part rate but sought "no formal action." Flexibility was obviously the key here.[5]

By the spring of 1901, the last rate war in Denver was under way. Small rate cuts could not turn aside the passions of the day or Charles LaCombe's enthusiasm for a new electric plant under his direction (which the city had approved on March 30, 1900, as described in the previous chapter). In the near term, Doherty predicted an expensive contest for a limited amount of business. In February, directors had voted not to pay a semi-annual dividend. Earnings were adequate to make the usual payment, they were told, but "considerable sums of money will be needed in the near future for improvements and extensions." On March 31, Doherty secured permission from the board to "advise all consumers to be prepared to get full benefit of a long continued rate war."[6]

Doherty did not simply cut rates again. Instead, on April 1, he announced implementation of the three-part rate originally sketched in 1894. First, households would pay a flat fee of $12 a year for their lighting hookup ($24 a year for power customers); the second part of the rate was Doherty's readiness-to-serve charge, another fixed fee based on the potential demand of each customer and hence a rough measure of the cost to the company of installing expensive equipment

5. DG&E Meetings, p. 104.

6. Ibid., pp. 124, 127. Details of the LaCombe contract with the city are described in several newspaper clippings in Hale Papers, Box 9, File 549.

and holding it in readiness. Doherty set that charge at $1.80 a year for each light bulb connected to his lines, and $24 a year for each horsepower. Actual consumption determined the third portion of the monthly bill, with charges of 5 cents per kilowatt-hour for incandescent lighting, and 3 cents for each kilowatt-hour used in an electric motor. As part of a "move to meet threatened competition," customers accepting the three-part rate would actually be charged only 50 percent of the stated rate. Promises of free light bulbs, similar treatment of customers "regardless of location," and the pledge that "incandescent and power circuits will be energized continuously throughout the 24 hours of the day" were among the several advantages that competition also produced in Denver.[7]

Doherty had calculated his three-part rate and other incentives with a view toward politics, geography, and the economics of utility operations, not just another rate war. Doherty believed that fixing portions of the rate would appeal to higher-volume customers, producing a steady and slowly growing demand. If rates were made particularly attractive to larger users, operators of electric plants in factories and downtown buildings might convert to a central hookup. In contrast, occasional users, not profitable because the cost of serving them was greater than the income generated, could be "weed[ed] out." During the rate war, Doherty encouraged low-volume users to switch to LaCombe, deferring the readiness-to-serve charge for several months "to enable them to obtain service from our competitors." Doherty "let . . . rivals have the cheap trade," the compiler of his papers later reported, and cultivated the "select trade" for his own business.[8]

Equally important was Doherty's creation of a rate structure that appealed to the politics as well as the economics of wealthier households. They were articulate, organized, and knowledgeable about public policy in areas that affected the cost and delivery of urban services such as trolley, water, gas, and electricity. In addition to favorable rates, the select trade was to be cultivated with around-the-clock service, which was being described as energizing the circuits on a continuous basis. Still another lure was the promise of similar treatment "regardless of . . . location." In 1902, only the select trade and operators of their own electric plants stood to benefit from a longer period of

7. DG&E Meetings, pp. 125–127; *Denver Times*, April 5, 1901 (microfilm), Denver Public Library.

8. Marston, *Principles and Ideas*, pp. 10, 283–284; *Denver Times*, April 5, 1901 (advertisement, on microfilm), Denver Public Library. For Samuel Insull and the politics of rate-making in Chicago, see Harold L. Platt, *The Electric City: Energy and the Growth of the Chicago Area, 1880–1930* (Chicago: University of Chicago Press, 1991), pp. 86–87.

service, and by definition none but the select trade, locating in areas distant from the central business district, stood to benefit from service that did not penalize residents of the city's cool green rim. The overall intent of Doherty's three-part rate was to co-opt members of the select trade in urban politics. A uniform rate, Doherty told utility executives in 1904, "is absolutely inequitable, . . . compels us to sell the product at a price . . . much in excess of the manufacturing cost . . . , [and] exposes us to all sorts of Populistic legislation."[9]

By May 1902, this last round of competition had run its course. Citing high costs for such items as copper and poles, holders of a majority of the shares in the LaCombe company sold out to members of a group associated with McMillin and Doherty. LaCombe refused to continue as president of the company. An account in a local newspaper described him as in a state of "mental and physical depression" brought about by criticism of his management and the turn of events. Doherty had the city to himself again, though not many customers, and a contract to light the city's streets. But the rate war had also depleted Doherty's treasury, according to a reporter's account, making it difficult to raise funds for expansion at reasonable rates. In addition, debts incurred in the consolidations of the nineteenth century were falling due. Doherty preferred bankruptcy and was able to arrange for a favorably inclined judge to name him as receiver. LaCombe attempted suicide, cutting his left wrist with a shaving razor on Independence Day.[10]

Rates and service changes were not Doherty's entire answer to competition and politics, or even the most important. Doherty also evolved a marketing strategy aimed at converting advantageous rates and service into solid increases in the consumption of gas and electricity. As early as November 1900, Doherty had hired a manager of new business at the whopping annual salary of $3,600 plus membership in two clubs. Doherty's reasoning was identical to the logic of load-building that was beginning to emerge in the electric and gas industry nationwide. "Opportunities for saving in manufacturing [gas and electricity] are almost insignificant," he wrote executives in the McMillin group during the LaCombe rate war in 1901, "compared with the opportunity of earning money by increasing the output." It appeared to him that "money . . . expended on . . . campaigning can be made

9. Henry L. Doherty, "Comment," *Proceedings of the American Gas Light Association*, vol. 21 (N.p., 1904), pp. 259–260.

10. Newspaper clippings, Hale Papers; Charles A. Frueauff, "History of the Company Gas Properties," May 1, 1905, SHSC; DG&E Meetings, pp. 202–203.

to pay 100 percent on the investment."[11] By 1905, he had assembled a large sales force and ordered members to canvass every home and business in the city for new business on his three-part rate. Marketing was one tool managers such as Doherty used to control the effects of public policy and rapid out-migration.

Doherty was not content with the idea that stable rates and a growing volume were the path to stability in the urban political arena. Members of Doherty's sales force were also expected to cultivate goodwill for the company. "If a customer of a dry goods house becomes offended at one store," Doherty told a meeting of office employees on September 7, 1904, "they can easily go to another." But in the gas and electric business, he added, "they are compelled to patronize us, and they resent that." Rude treatment of customers accumulated in the form of bad feeling toward the company. Because office and sales personnel had contact each day with consumers, it appeared possible to alter perceptions of the company. "It is possible to treat a customer who thinks he has a great grievance, so that he will at least feel as friendly toward the company as he did before." Doherty concluded, "We want the friendship of the public."[12]

Cultivation of friendly feelings offered an entryway into sales increases as well as an exit from negotiations in the formal political arena regarding rates and service. "We are now out of politics," one of Doherty's senior executives announced at a meeting of sales personnel on June 26, 1906, "and we want the public to feel that our connection with politics is entirely severed." On May 15, 1906, Denver's voters had approved a new franchise for the Denver Gas & Electric Company. In return for a twenty-year franchise, Denver's politicians and consumers were promised a payment annually of $50,000 to the city, lower rates, installation of poles painted in a manner approved by the city's Art Commission, improved street lighting, and extensions of service into fast-growing areas recently incorporated into the city. By June 18, a company official reported to participants at a sales meeting that crews were "adding to the West Side station" and "building a large distribution feeder out 23rd Street and Park Avenue."[13]

In reality, Doherty and his associates were never "out of politics." Indeed, political leaders had created the framework within which

11. DG&E Meetings, p. 123; Marston, *Principles and Ideas,* p. 154.

12. "Minutes of Meeting of Office Association," September 7, 1904, Munroe Files.

13. *DG&E Bulletin* 1 (June 1906), 11, 13, Munroe Files; Clyde Lyndon King, *The History of the Government of Denver . . .* (Denver: Fisher Book Company, 1911), pp. 277–279.

Doherty shaped his program of rates, expanded service, and sales promotions. In addition, Doherty and his senior officials still dealt routinely with politicians on such matters as rates and extensions of service, including installation of a large number of street lamps for which a Doherty executive as of June 18 was "waiting for the city officials to give us the word." Beyond that point, rate-making, salesmanship, extensions of service, and securing a positive opinion of the company and its employees were part of a larger contest for the ability to allocate electric and gas resources, dispense prestige, and shape a business environment favorable to stable and predictable growth. Politics, formal and informal, remained important, especially the politics of boosting demand for electric and gas service and cultivating a positive image for the company among local residents. They still went hand in hand. "Tomorrow morning," one of Doherty's executives told sales personnel on June 14, 1906, "I would like to have each man prepared to offer some suggestion as to how we can increase the popularity of this company, and how we can at the same time increase our revenue through the sale of gas and electricity." Thomas Hughes, following Isaiah Berlin, describes Samuel Insull as a hedgehog—someone "who relate[s] everything to a single central vision."[14] Doherty was a hedgehog too. Both Doherty and Insull were, in fact, urban hedgehogs.

Although Doherty had escaped the day-to-day problems of dealing with rate wars and the threat of municipal takeover, he had to adapt operations to rapid out-migration of population and to an evolving consumer politics. Curiously, not even the monopoly status granted by city and state governments guaranteed success, or even survival, for the gas and electric firms operating in Kansas City. Only after World War I did a new management group at the electric company in Kansas City succeed in stabilizing operations. Executives of Kansas City's gas company, known since the rate war of 1896 as the "old company," resisted compliance with public policy and urban change until the mid-1920s (at which point Doherty purchased the company). The essence of the problem at both firms was an inability to achieve an acceptable balance among new technologies, economies of scale, urban growth and change, and public policy.

On January 9, 1900, J. Ogden Armour and several associates assumed direction of the Kansas City Electric Light Company (formerly Kawsmouth), replacing Edwin Weeks and a board of directors com-

14. *DG&E Bulletin* 1 (June 1906), 9, 11; Hughes, *Networks of Power,* p. 18.

prised principally of local entrepreneurs. Armour was the son of Philip Armour, founder of Armour & Company, the large packinghouse. J. Ogden Armour already held a controlling interest in Kansas City's Metropolitan Street Railway Company and was in the process of consolidating the remaining trolley companies. Weeks had contended that massive investments were required to serve the city with modern and efficient equipment, and during the next five years Armour spent more than $3.3 million to build and modernize his generation and distribution system. In addition, an engineering journal reported in May 1903 that the holding company, the Kansas City Railway & Light Company, was capitalized at $25 million.[15] At first glance, the backgrounds of the new operators, along with the vast outlay of funds as part of a holding company, represented the ascendance of national finance and the dilution of technique and locally oriented political action. Actually, neither politics nor technique was lost, and finance was not even supreme; Armour and his associates were simply unable to achieve an effective coordination among them.

The integration of trolley and electrical systems was a popular idea among utility operators in North American cities. One component of Samuel Insull's success in developing the Commonwealth Edison Company in Chicago was securing contracts to supply current to local streetcar companies. By 1910, according to Hughes, leaders of the nation's electric industry considered Insull's Chicago operations "to be the world's greatest." Insull's reasoning (and Armour's was no doubt similar) was that by combining his traction, power, and lighting loads in a large plant it was possible to enhance plant utilization and to lower unit costs. The logic was simple. Demand for current to service factory power, trolleys, and street and household lighting peaked at different hours of the day. Consequently, the company was not required to purchase additional and costly steam turbines and high-capacity generators for occasional peak loads. In turn, higher-volume consumers, especially factory operators, were to be lured by discounts into making additional purchases of current, thus filling in the troughs on the load curve. Between 1905 and 1910, Armour spent another $2.1 million to enlarge and upgrade generation and distribution systems. Similar to central electric operations throughout North

15. Edwin D. Shutt, "The Saga of the Armour Family in Kansas City, 1870–1900," *Heritage of the Great Plains* 23 (Fall 1990), 25–42; Kansas City Power & Light history, pp. 23, 26, in K.C.P&L; [Philip J. Kealy], *Report on the Fair Value of the Property of the Kansas City Electric Light Company and Subsidiary Companies as of February 1, 1914* (Kansas City, Mo.: Smith-Grieves Company, 1914), p. 103; "Kansas City Consolidation," *Electrical World and Engineer* 41 (May 30, 1903), 943.

America, one power plant manufactured electricity for consumers throughout the city and surrounding area, including many of the large factories located across the river in Kansas City, Kansas. In turn, rotary converters, frequency changers, transformers, and other modern equipment allowed Armour to produce both direct current for streetcars and other high-torque motors as well as alternating current for illuminating purposes.[16]

The technical features of Armour's operations in Kansas City replicated those appearing in Denver, Chicago, and other large cities. Yet Armour failed to create an organization capable of managing those investments. Monopoly proved of little benefit in the long run when operations had to adapt to a rapidly changing city. A key problem was Armour's early failure to shape and train a sales force that could deliver an adequate number of customers and provide regular input about customers' changing needs.[17]

In 1905, Armour created a sales department to foster increased consumption of electricity. Initially, management and organization of the sales force duplicated promotional efforts taking place under the direction of electrical operators in other cities. Armour placed the sales force and overall management of the electric company in the hands of R. E. Richardson, an executive who had joined the company around 1900. In turn, Richardson assigned members of the sales force to territories, insisting on door-to-door efforts similar to those Doherty had initiated in Denver. Several representatives concentrated in Kansas City, Kansas, where manufacturers seeking access to slaughterhouses, railroad yards, or the Kansas River had located. Another group of representatives specialized in window lighting, which was popular among electrical operators in other cities because window lights burned long hours at night, when industrial, trolley, and household demand slackened. In 1906, moreover, Richardson assigned a salesman to door-to-door marketing of electric irons; and in 1907, the company opened a store selling toasters, chafing dishes, and other appliances. Yet Richardson employed only 11 sales representatives in all, assigning 4 to work in Kansas City, Kansas, and directing the remaining 7 to promote electrical merchandise in Kansas City, Mis-

16. Hughes, *Networks of Power,* pp. 204, 216–223; [Kealy], *Report on the Fair Value,* p. 103; "Crane Equipment of the Missouri River Power Station, Kansas City, Mo.," *Electrical World and Engineer* 44 (October 29, 1904), 741.

17. According to Alfred D. Chandler Jr., *The Visible Hand: The Managerial Revolution in American Business* (Cambridge: Belknap Press of Harvard University Press, 1977), p. 399, Armour was a member of a group of packers who "paid little attention to strategic planning and the long-term allocation of resources."

souri. In 1905, by contrast, Doherty employed more than 40 representatives in Denver, a city smaller than Kansas City; and by 1909, Samuel Insull employed 125 in Chicago. Not only did operators in Denver employ a larger sales staff, it was also possible to deploy them as a group with a single-minded focus on one product—electric irons, for instance.[18]

In contrast to regular practice in the electric and gas industry, Richardson's company chose not to advertise in the city's newspapers. Kansas City was not large enough to justify placing advertisements in every newspaper, Richardson contended in 1905, and selecting only a few papers would create "hard feeling." During the three-year period 1906–1908, then, outlays for advertising totaled only $11,000, and the salaries of sales personnel for residential and commercial lighting as well as power amounted to nearly $54,000. As a point of comparison, in 1910 alone Doherty spent more than $18,000 to advertise both gas and electricity and paid salaries and commissions totaling nearly $45,000 to members of his sales force in the smaller city of Denver.[19]

Richardson and Armour spent comparatively little on advertising, and nothing on educating their employees in the strategies and details of company operations. Indeed, Richardson and Armour reversed a tradition established by Weeks. In 1887, Weeks had organized the Gramme Society of Kansas City, so called after Belgian electrical inventor and engineer Zénobe Théophile Gramme (1826–1901). The company donated books on electricity, as well as a room for meetings, which were held twice a month. Beginning in 1894, moreover, Weeks and other leaders of the society organized courses in physics, electricity, and arithmetic. Instructors, who were also employees, included a draftsman and dynamo tender, suggesting a practical turn in course contents. In 1897, the courses were discontinued, leaving a committee on education that included Weeks as a member, but in 1900 Armour discontinued that committee. In 1911, Armour and Richardson reinstated instructional activities, but that phase lasted only two years, at which point the company turned over the courses to a local trade

18. "Building up Central Station Business at Kansas City," *Electrical World and Engineer* 46 (November 25, 1905), 894; Hughes, *Networks of Power,* p. 224; Kansas City, Missouri, Public Utilities Commission, *Report of the Public Utilities Commission on the Financial Conditions of the Kansas City Electric Light Company and Subsidiary Companies* (1909), Exhibit C; Kansas City Power & Light history, p. 27.

19. [Kealy], *Report on the Fair Value,* p. 81; Denver Gas & Electric Light Company, "Electrical Manufacturing and Expense Reports, 1910" (Document 262–1) and "Gas Manufacturing and Expense Reports, 1910" (Document 262–2), both in PSCC.

school.[20] By way of contrast, gas and electric company employees in Denver were acquiring increased sophistication in making connections between rates, politics, and urban growth. After 1900, however, employees of the Kansas City Electric Light Company lacked frameworks in common for connecting political and geographic change in their city with equally rapid growth and change in the methods of electrical generation, transmission, and distribution.

Nor did Armour and Richardson create a rate structure that satisfied consumer politics. As late as 1900, the original owners of Kansas City's electric company had a published rate of 20 cents per kwh. Within a few years, Richardson reduced rates to 10 cents per kwh for household users, mostly for lighting purposes, and offered still-lower rates to industrial and commercial accounts that promised long hours and heavy demand. In 1905, Richardson was able to declare that residents of the Kansas City area had once "looked upon electric light as a luxury" but that lower rates and promotional activities had already "brought them to consider it a necessity." During the next four years, Richardson continued to emphasize that his rates were among the lowest in the nation. By 1909, however, the precise amount that Richardson was charging customers was only one of several factors at issue between him and local opponents.[21] Increasingly, ideas about equity and authority animated utility politics.

Between 1900 and 1910, the urge to regulate companies supplying light, heat, and transportation was reaching its peak in North American cities, including Kansas City. In 1907, a group described by local businessman and reform enthusiast Jacob A. Harzfeld as "public spirited citizens" secured approval from the Missouri legislature for municipal authorities to fix rates charged by local utility operators. In May 1908, the mayor appointed a commission with authority to investigate rates and make recommendations to him and to members of the upper and lower houses of council. Preliminary investigation by the commission, of which Harzfeld was now a member, showed that rates charged in Kansas City varied between 3 cents and 10 cents per kwh. Worse yet, "residents operating in the same line of business upon the same street and using approximately the same amount of current

20. F. Y. Hedley, "Gramme Society," in Howard L. Conrad, *Encyclopedia of the History of Missouri* (New York: Southern History Company, 1901), 3:83; Hughes, *Networks of Power*, p. 87; "History of Educational Activities of the Kansas City Power and Light Company," *The Tie* 6 (April 1926), 6–7.

21. "Building up Central Station Business at Kansas City," p. 894; Kansas City Public Utilities Commission, *Report on the Financial Conditions of Kansas City Electric Light Company* (1909), pp. 33–34, 38–39.

were receiving different rates." So unfair had the situation become, added Harzfeld, that "customers' meters were inspected by an officer whose salary was paid by the Company, whose assistant was a direct employee of the Company," and "the instruments used were instruments belonging to the Company, the City owning none of its own."[22]

Of course, the ideologies of the day shaped the rhetoric of reformers such as Harzfeld, just as a variation on those ideologies shaped the observations of leaders in the electric business, such as Richardson. By 1910 in Kansas City, discussions of rates, whether by Harzfeld or Richardson, expressed a more fundamental disagreement. The core of the matter was that, formerly, negotiations about matters such as rates and service had taken place on an individual basis between company and customer. But private negotiations no longer appeared satisfactory, especially among larger businesses and wealthier households that were increasingly reliant on large firms such as the gas and electric company to supply the everyday necessities of industrial, urban, and family life. Doherty had shaped his readiness-to-serve rate for members of this group. In reality, then, the commission was one of the devices shaped with increasing frequency after 1900 by politicians who were seeking to create a formal arena in which to conduct negotiations that had failed in private settings.[23]

Richardson proved to be a miserable negotiator. In mid-1909, he testified before the city's newly established commission regarding rates charged industrial and household customers. During the period 1900–1903, he had "sold current at one-third of what it cost us." Similar to Insull and Doherty, Richardson figured that installation of modern steam turbines and high-power generators would eventually lower production expenses. Rates had been adjusted during the next few years. Yet numerous groups of customers paying different rates had been created in the process, which was exactly the conclusion reached by proponents of rate regulation, such as Harzfeld. As Richardson understood the problem, however, industrial and commercial customers often contracted for large quantities of power and lower rates, but then failed to consume the power for which they had contracted. Different rates for identical levels of consumption appeared

22. Jacob A. Harzfeld, "The Utilities Commission of Kansas City, Missouri," in Clyde Lyndon King, ed., *The Regulation of Municipal Utilities* (New York: D. Appleton & Company, 1912), pp. 219, 221, 226–227.

23. Ibid., pp. 219–223, 226–228; KCP&L history, pp. 21–28; Carrie Westlake Whitney, *Kansas City, Missouri: Its History and Its People* (Chicago: S. J. Clarke Publishing Company, 1908), pp. 276–279; K. Austin Kerr, *American Railroad Politics, 1914–1920: Rates, Wages, and Efficiency* (Pittsburgh: University of Pittsburgh Press, 1968), pp. 1–2.

inevitable. Overall, Richardson was able to conclude only that "there is not a larger question before the engineering fraternity today that is as complicated . . . as . . . the rate question."[24]

Richardson also failed to comprehend the politics of rate-making at the household level. He had fixed a minimum charge of $1 per month and a uniform rate of 10 cents per kilowatt-hour on household customers. Nonetheless, he believed that setting rates based on the potential demand for current (which was one portion of the system imposed by Doherty in Denver) was "absolutely impossible [even] if I had a corps of 10,000 men." Customers often installed additional lighting after signing a contract, he told members of the city commission, leaving the company "absolutely at the mercy of the public." In order to be certain that customers did not cheat, the company would have to "inspect 1,300 or 1,400 houses a day." What did appear desirable was Doherty's method of charging both a minimum per month plus a second minimum based on potential demand, and then they "sell you the current for 5 cents." He was willing to develop that system for Kansas City, he told members of the commission, provided "you would let us." If a more equitable method of fixing rates were developed, he concluded, "it would relieve us of unlimited work [and] all sorts of trouble."[25]

By about 1909, Richardson had reached a juncture in the politics of rate-making. He could reduce the company's flat rate, but with no certainty as to what that course would purchase in terms of improved demand or political favor. Alternatively, he could attempt to secure a change in his relationships with city officials and politically astute consumers. Doherty had succeeded with that strategy in Denver. Yet in his public conversation with members of the commission, Richardson spoke of the Doherty rate in terms of being "relieve[d] . . . of all sorts of trouble." Whether those in political trouble are relieved by their opponents is a debatable point, but Richardson's choice of words suggests a perception of the political universe as constraining. After 1909, only squabbling among Kansas City's politicians and a franchise granted in perpetuity to the founders of the Kawsmouth company in 1881 allowed Richardson's operation to continue a while longer without direct interference.

Between 1900 and 1909, Richardson and his superiors had not developed the skills and savvy that were requisite to success in a turbu-

24. Kansas City Public Utilities Commission, *Report on the Financial Conditions of Kansas City Electric Light Company,* pp. 23–28.

25. Ibid., pp. 27–33.

lent urban arena. As political noise escalated and business declined, Richardson and Armour chose not to make a special effort to win the allegiance of political and business leaders in the Kansas City region. After 1909, neither extensions of service to outlying areas nor a vast increase in the number of customers were relevant facts in the region's politics.[26]

In the absence of social and political know-how, Richardson and the Kansas City Electric Light Company were vulnerable to every disturbance in the city's political arena. Connection with Armour's trolley operations brought about bankruptcy and reorganization at the electric company. Similar to Richardson, managers of the trolley company had failed to adjust operations to changes in the geography and politics of urban transportation.

Between 1900 and 1915 or so, the politics of public transport in Kansas City evolved along lines similar to those in larger cities in North America, such as Chicago, Toronto, Boston, and New York. Beginning in 1902, the "Met," a popular name in Kansas City for Armour's Metropolitan Street Railway Company, operated under a "Peace Agreement" with city officials. Executives of the Met agreed to universal transfers and to a 5-cent fare throughout the city as well as south to Swope Park, a popular recreational area. Nor could the company permit service frequencies to drop below agreed-upon levels. The peace agreement also required payment to the city of 8 percent of gross proceeds in lieu of other taxes along with construction each year of two miles of double track. As in many cities, transit officials had accepted regulation, service improvements, and cash payments as the price for long-term stability in the political arena and the continuing opportunity to earn hefty profits.[27]

26. *Kansas City Times*, December 12, 1912, and December 29, 1913; *Kansas City Star*, April 9, 1916, and February 19, 1922, all in clipping files, Kansas City Public Library; "Constructions," *Electrical World*, January 25, 1913, 219; "Appraisal of the Kansas City Electric Light Company and Subsidiary Companies of Kansas City, Missouri" (typescript, 1915), pp. 38–40, K.C.P&L. For the state of political alliances in Kansas City, see Lyle W. Dorsett, *The Pendergast Machine* (New York: Oxford University Press, 1968), pp. 50–61. Between 1911 and 1914, the number of meters installed increased from 19,000 to 36,000, and kilowatt-hours sold increased from 31 million to 69 million. See *Kansas City Star*, February 19, 1922, and compare with Kansas City Railway & Light Company and Subsidiary Companies, *Annual Report for the Fiscal Year Ended May 31, 1911*, pp. 10–11.

27. Dorsett, *The Pendergast Machine*, pp. 59–60; Whitney, *Kansas City, Missouri*, p. 12; Paul Barrett, *The Automobile and Urban Transit: The Formation of Public Policy in Chicago, 1900–1930* (Philadelphia: Temple University Press, 1983), pp. 38–45; Charles W. Cheape, *Moving the Masses: Urban Public Transit in New York, Boston, and Philadelphia* (Cambridge: Harvard University Press, 1980), pp. 208–219; Armstrong and Nelles,

In retrospect, transit operators made bad bargains for themselves, setting in place a system that restricted revenues and boosted expenses. During the 1920s and 1930s, these agreements and declining patronage drove their successors into reorganization; after World War II, these same arrangements and patronage that declined further drove them out of business. In Kansas City, moreover, the peace agreement lasted only until June 1911, at which point directors of Armour's holding company filed for bankruptcy on behalf of the Met; and in November 1914, directors of the holding company tossed the electric company into the judicial and political deliberations.

Early in 1916, a federal judge handling the bankruptcy proceedings separated the trolley and electric companies and launched each as independent firms. Because in his view the future of the trolley company held out the prospect of better earnings than those of the electric company, he assigned most of the generating capacity in the city to the newly independent trolley company. The new electric company would purchase supplementary current from the trolley firm.[28] Altogether, the judge's decision to award custody of most of the generating capacity to the Met suggests a continuing perception among the city's leadership that Richardson had failed to satisfy popular passions, consumer politics, or public policy.

In the absence of stable relationships in the political and then in the judicial arena, funds for improving electrical as well as trolley service had been unavailable. By 1917, as production for World War I encouraged an increase in transit ridership and a rise in the demand for light and power, the Met lacked a sufficient number of cars and the new electric company's generating capacity was inadequate. In August 1917, about all that executives of the reorganized electric company could promise was that "the condition will improve." Years later, the new president of the company reported that the exigencies of war, along with "lack of attention to plants due to receivership and reorganization [had] so overwhelmed the poorly maintained machinery that it became necessary to discontinue service at certain hours of the day to the industries depending on the light company for

---

Monopoly's Moment, pp. 264–266; Harzfeld, "The Utilities Commission of Kansas City, Missouri," in King, *Regulation of Municipal Utilities,* p. 224; Olson, *Kansas City Power & Light,* p. 14.

28. Joseph F. Porter, "The History of the Electrical Industry in Kansas City and Vicinity" (thesis, Iowa State College, 1934), pp. 17–18, 22. For the tone of political discourse regarding purchase by the electric company of current from the trolley company, see *Kansas City Times,* December 12, 1912, and December 29, 1913. Leaders of the electric company could not extricate themselves from these perceptions.

power."[29] It would take several years, however, for new operators to bring rates, organization, and technologies into conformance with the political geography of Kansas City.

Gas supplies in Kansas City virtually collapsed during World War I, leading to cold dinners and dark and chilly rooms during the war years. But this dreary situation was not an inevitable result of the war. A series of choices made by pipeline operators, politicians, and executives of the old gas company actually created the shortages. During a period that extended back to the rate war of 1896, negotiations about rates and about service to the fast-growing neighborhoods on the city's periphery dominated the politics of gas in Kansas City.

Beginning in 1902, the presence of vast deposits of natural gas in southeastern Kansas encouraged formation of companies with the goal of building a pipeline to Kansas City. A journalist later reported that the delivery of natural gas to Kansas City had been "championed by a set of men who at the time had neither gas nor mains." Among most of these speculators, two ideas appeared paramount. First, each hoped to secure a contract with the old company (or its successor) and with gas companies in other municipalities located along a route from southeastern Kansas to Kansas City. Second, award of that contract would serve as collateral for the sale of bonds and stocks that were going to be needed to finance so vast an undertaking. In brief, a small investment protected by contracts might yield a fabulous return. Natural gas located in the mid-continent region of Kansas and Oklahoma was plentiful and cheap and would replace water gas, which the old company had been selling since 1897 at the now-exorbitant price of $1 per thousand.[30]

With the prospect that a pipeline company would soon provide lots of low-priced natural gas, members of the city council in Kansas City began to debate lower rates and the award of a new franchise (to replace the old company, whose franchise expired in 1906). By late

29. Clyde Taylor, "Facts Concerning the Street Railway and Power Situation in Kansas City Submitted at a Meeting of the Business Men's Committee of the Chamber of Commerce, Tuesday, August 28, 1917" (n.p., n.d), p. 5, in Kansas City Public Library; and Porter, "History of the Electrical Industry in Kansas City," pp. 20–22. For an account of public policy and shortages of gas and other fuels during World War I, see John G. Clark, *Energy and the Federal Government: Fossil Fuel Policies, 1900–1946* (Urbana: University of Illinois Press, 1987), pp. 81–109.

30. *Kansas City Star,* October 1, 1904; *Kansas City Star,* December 13, 1904; *Kansas City Journal,* September 28, 1906—all in clipping files, Kansas City Public Library.

1904, members of city council crystallized the presumably compatible passions for cheap gas and a lucrative franchise into legislation. A rate of 40 cents per thousand cubic feet (mcf) was the centerpiece of this bill, which amounted to a proposed reduction from the existing price of 60 percent. Because natural gas contained approximately double the caloric content of manufactured gas, the rate decrease was actually far greater. But not even a new product, such as natural gas, and a sharp drop in rates could emancipate the supplier from the history of utility politics in the city and nation. Leaders of the old company emerged as the likely franchisee, which reflected no particular clout at City Hall, but rather the fact that they had pipes in the ground and a demonstrated ability to complete a large project quickly. On December 15, however, Mayor J. H. Neff vetoed the ordinance, citing violations of the city's charter and his determination to secure an option to purchase the plant.[31]

Time and economics favored politicians as well as residents along the city's periphery. By mid-1906, the pipeline from southeastern Kansas had nearly reached Kansas City. Expenses of construction, including litigation costs and the repair of sabotage by residents along the right-of-way, had pushed expenses much higher than anticipated. Even more, by July 1906, the new mayor of Kansas City, H. M. Beardsley, threatened award of the local distribution franchise to a group of business leaders whose identity would remain secret. "I have seen every group of men who have come out heretofore bought off or frightened off by the power of the U.G.I.," he observed, and "I don't want these men subjected to this pressure until they are so deep in they can't get out."[32]

The mayor's threats worked. In late September 1906, a deal with leaders of the old company to sell natural gas in Kansas City was set. Now, rates for householders were fixed at 25 cents per thousand, with an increase to 30 cents permitted in ten years, rather than in the five-year period for which company executives had hoped. Large consumers of gas, such as factory operators, were to receive an even lower rate, which in practice amounted to 10 cents per thousand. Consumer politics won another victory in a liberalization of the number of customers required per gas line. Earlier, persons residing in sparsely settled sections of the city had to present a petition from six house-

31. *Kansas City Times,* October 1, 1904; *Kansas City Star,* December 13, 1904, and December 21, 1904 (both items not previously cited), in clipping files, Kansas City Public Library.

32. "The Mayor on Gas," *Kansas City Times,* July 9, 1906 (typescript), in Snyder Collection.

holds spaced within 400 feet of each other in order to secure gas service. The ordinance of 1906 reduced the figure to three within 200 feet. In effect, only three households clustered near each other on a block in one of the distant subdivisions could insist on gas service, at once both increasing the expenses of the old company and no doubt adding to the potential of developers in the far-off districts to boost the sale and rental of houses and apartments. "The city administration," an executive of the pipeline company contended in 1913 as supplies began to run low, had taken "advantage of the situation to drive a hard bargain, almost confiscatory in nature."[33]

Arrival of natural gas in Kansas City initiated the second boom in gas distribution and consumption. By late May 1907, the old company was supplying gas to 36,000 customers through more than 330 miles of underground pipe. Gross revenues of the old company for the period January 1–June 30, 1907, jumped above $1 million, reported a local historian, more than doubling the income produced during the last six months of 1906. This "advance," as she labeled it, "was due to the general substitution of gas for coal as a fuel for cooking and heating in dwellings and very largely in hotels and office buildings and even to a considerable extent in manufacturing establishments." By 1910, some 52,000 meters were in service, and the old company's revenues exceeded $2 million. During 1910 and 1911, as the "advance" continued, consumption of gas for all purposes exceeded 17.4 million cubic feet. In 1911, the second gas boom peaked. During 1915 and 1916, consumption declined to 11.5 million cubic feet, a drop of 33 percent.[34]

The earliest signs of problems in delivering adequate supplies of natural gas had appeared during the winter of 1910. Although "last winter the service was better than in the past," the city's gas inspector

33. Kansas City, Missouri, Public Utilities Commission, *Laws, Ordinances, and Permits Dealing with Rights, Privileges, and Franchises of Public Service Corporations in Kansas City, Missouri* (1912), pp. 134, 143–158; *Kansas City Journal,* September 28, 1906, in clipping files, Kansas City Public Library; "The A.B.C. of the Gas Situation in Kansas City," c. February 1913, in Snyder Collection. For the boom and decline in natural gas and associated industries, see John G. Clark, *Towns and Minerals in Southeastern Kansas: A Study in Regional Industrialization, 1890–1930* (Lawrence: State Geological Survey of Kansas, 1970).

34. *Kansas City Star,* May 22, 1907, clipping files, Kansas City Public Library; Whitney, *Kansas City, Missouri,* pp. 275–276; "Kansas City Gas Company: Gas Consumed from January 1910 to July 1923, Inclusive" (typescript, probably prepared by Robert M. Snyder Jr.), Snyder Collection; Public Utilities Commission, "To the Public," in Robert M. Goodnow, comp., *Information for the Natural Gas Consumers of Kansas City, Missouri* (Kansas City, Mo.: Gas Inspector's Department, 1911), p. 3.

reported in 1911, "some interruptions did occur," and he recommended "that you have some coal or wood on hand to fall back upon." In January 1912, the president of the pipeline company that brought gas to the old company from southeastern Kansas sought permission from members of the Public Utilities Commission of Kansas to discontinue the sale of natural gas at 10 cents per thousand to operators of smelters and other factories who burned it in place of coal as an inexpensive boiler fuel. "Natural gas is preeminently a domestic fuel," he contended, "and should never be prostituted to a manufacturing or gross use." At the present rate of gas consumption, he added, the company's pipes and equipment would soon "be a mere mass of worthless junk."[35]

By mid-1912, however, the pipeline company was in default on its bonds and in receivership. The federal judge assigned to the case contemplated an order requiring the company to charge a higher rate. In 1913, the president of the pipeline company asked political leaders in Kansas City (and other municipalities along the line) to approve an increase in the household rate to 35 cents per thousand. In bringing natural gas to the city in 1906 at 25 cents, he asserted, the company had "made a more sweeping reduction in one of the necessities of life than has ever been given the city . . . , and in return the people of Kansas City should reciprocate." Because city officials lacked incentive to boost rates, they remained at the 25-cent level until 1916, when the city ordinance permitted an increase to 30 cents. In November 1916, Henry Doherty, quick to recognize a potential bargain, purchased the company in receivership.[36] Public policy, then, had fixed the framework for elimination of the pipeline company's original stockholders and managers.

Operators of the old gas company in Kansas City were faring no better. By 1913, with the pipeline company in receivership, leaders of the old company remained as principals in an adversarial relationship with local politicians that extended back nearly two decades. Rates were going to remain as they were. Politicians also expected uniform distribution of gas service throughout the city. In fact, be-

35. "Some Suggestions Regarding the Use of Natural Gas," in Goodnow, *Information for Natural Gas Consumers,* p. 9; "History and Information Concerning the Gas Service Company, 1925–1957" (typescript, c. 1958), files of the Gas Service Company, Kansas City, Missouri; President, Kansas Natural Gas Company, to Public Utilities Commission of Kansas, January 9, 1912, in Snyder Collection.

36. Notice of Default Served February 1, 1913; newspaper clippings, c. July 1912; Report by Moore, Leonard & Lynch, November 11, 1916—all in Snyder Collection; "A.B.C. of the Gas Situation"; Clark, *Towns and Minerals,* pp. 69–70.

tween December 1909 and February 28, 1916, several thousand Kansas City residents signed petitions requesting gas service. The city's mayors and members of both houses responded with nearly 500 ordinances ordering officers of the old company to extend lines. Yet according to Harzfeld, as late as 1911 leaders of the old company had "refused, whenever possible, to extend gas mains unless they saw a profit in those mains."[37]

By 1916, political leaders could select from among four alternatives in securing gas for city residents. The first alternative was to permit the old company to raise rates. The company's executives were insisting that higher rates were requisite to improved service. But city officials were already disputing the pipeline company's authority to raise rates. In January 1916, receivers for the pipeline company had brought suit seeking a rate of 74 cents to replace the current rate of 30 cents. In September 1916, Harzfeld, now the city's attorney, advised members of council that the "conduct" of the pipeline company had been "indefensible" and that "Kansas City, Missouri, is resisting the establishment of those rates, is resisting the annulment of our franchise, and the annulment of the contracts."[38]

The second alternative was to locate a new supplier to replace the uncooperative old company. In January 1917, Henry Doherty, now the owner of the pipeline company, went to Kansas City to meet with local politicians. Following considerable success in Denver and several other cities with individual gas and electric firms, Doherty was turning increasingly to integrated operations. Rates and service were the questions at hand. Doherty announced his willingness "to spend whatever amounts are necessary to supply the gas required." The rate, he announced, did not need "to be greatly in excess of present rates" and was "not to exceed thirty-five cents." As late as January 1917, however, 35 cents probably looked too high to optimistic local politicians, though extant records fail to indicate the contents, if any, of subsequent negotiations with Doherty.[39]

37. Kansas City, Missouri, Records Management Section, Ordinance Record Locating Gas Mains (December 1909 through February 1916), vol. A141; Harzfeld, "Utilities Commission of Kansas City, Missouri," pp. 225–226; "History and Information Concerning the Gas Service Company, 1925–1927," p. 7; J[acob] A. Harzfeld to Members of Joint Committee of the Upper and Lower Houses of the Common Council Appointed to Investigate the Gas Situation, September 8, 1916 (microfilm), in Office of the Chief Clerk, Kansas City, Missouri. See also Platt, *The Electric City,* p. 128, for the "politics of confrontation in government-business relations" that characterized the gas business in Chicago.

38. Harzfeld to Members of Joint Committee to Investigate the Gas Situation.

39. *Kansas City Times,* January 4, 1917, in clipping files, Kansas City Public Library.

By late summer 1917, local politicians embarked on alternatives three and four. On August 28, members of both houses of City Council approved an ordinance directing city officials to acquire the facilities of the gas company. In the absence of funds to buy the old company, or the availability of an alternative operator such as Doherty, to whom management of its property could be entrusted, a local journalist interpreted the ordinance to mean that the city had simply "canceled the franchise." Realistically, about all that politicians could accomplish in the short run was to turn to the fourth and final alternative. They limited the consumption of gas. In November 1917, as the people of Kansas City faced another winter of flickering lights, chilly rooms, and uncooked meals, City Council ordered residents not to burn natural gas for space heating.[40]

During the course of several decades of dealing with gas and electric firms, city officials had acquired considerable skill in setting frameworks for utility operations. One result of public policy was that utility operators in Kansas City and Denver, as well as in cities throughout North America, had undertaken massive promotional programs aimed at boosting consumption and lowering unit costs. In Kansas City, however, public policy had a vindictive side as well. Politicians fixed prices and service requirements at levels below which managers of the electric company, the pipeline company, and the old company could operate.

Public policy had two sides. Politicians could shape the environment within which gas and electric firms operated. Yet politicians were unwilling to reshape that environment once management proved incapable of adapting operations to it. As late as September 1916, for example, the city attorney for Kansas City could still complain to members of Council that the pipeline company had "continue[d] to pay bondholders, both principal and interest, instead of investing for the needs of the consumers."[41] Habits of mind acquired in earlier contests were not easily shaken. Given at once the rates that seemed reasonable, the rapid out-migration of the population, and the consequent mandate to serve distant customers, it may have been the case that the situation for politicians and for leaders of Kansas City's gas and electric firms was virtually an impossible one. Public policy carried with it a momentum of its own.

40. *Kansas City Times,* August 28, 1917; *Kansas City Journal,* November 15, 1918, both in ibid.; Ordinance 31544, November 5–6, 1917 (microfilm), in Office of the Chief Clerk, Kansas City, Mo.

41. Harzfeld to Members of Joint Committee to Investigate the Gas Situation.

The politics of natural gas in Kansas City contained two final ironies. First, Robert M. Snyder, who had initiated the rate war of 1896, was also a large stockholder in the pipeline company. In 1912, his son, Robert M. Snyder Jr., attempted to block foreclosure. During 1916, the younger Snyder tried to prevent Doherty from purchasing control of the firm, but he was unable to marshal the financial, legal, and political resources required for these massive undertakings. Thereafter, the younger Snyder prepared charts of gas prices and consumption and wrote tendentious accounts of gas politics. In 1896 and again between 1912 and 1916, the Snyders senior and junior had stood on the unlucky side of public policy.[42]

A federal judge was responsible for the second irony in the city's gas politics. In September 1917, following lengthy and no doubt expensive litigation, the judge ordered receivers for the pipeline company to boost the price charged domestic customers for gas to 60 cents per thousand. In November 1918, the judge ordered the rate increased to 80 cents. External authority had altered what the city and gas company officials had been unable to alter since 1906. Only a federal appellate court or the U.S. Supreme Court could offer relief to politicians.[43] Remarkably, on March 17, 1919, the Court determined that a federal judge had no authority to fix rates inside of Kansas City. Nonetheless, the Court left standing the district court's order prescribing the rate of 80 cents for gas delivered to the city's edge; although the court had found for the autonomy of the city in fixing rates, the price of gas was going to remain at 80 cents.

The reality of the matter was that in both Kansas City and Denver, politicians and their policies were preeminent. In both cities, public policy had created a set of circumstances that suggested the logic of mass production and mass consumption. Public policy served as an external discipline to which operators of gas and electric firms had to adhere. Doherty shaped his rates to accommodate that policy and he shaped an organization that interpreted urban change with a view toward selling appliances to Denver's householders. Neither J. Ogden Armour nor executives of the pipeline company and the old company had built organizations that were adaptable to public policy. Not until

42. Robert M. Snyder Jr. to R. A. Long, May 28, 1913; R. M. Snyder Jr. to the *Star,* c. 1914; Robert M. Snyder Jr. to Stockholders, June 2, 1916; John C. Bartlett to Stockholders, August 17, 1916; [Robert M. Snyder Jr.], untitled and handwritten history of Kansas Gas Company, c. 1915—all in Snyder Collection.

43. *Kansas City Journal,* November 15, 1918, and May 21, 1920; *Kansas City Times,* April 10, 1919, and May 11, 1920; *Kansas City Star,* May 24, 1919; items dated April 10, 1919, May 24, 1919, and May 21, 1920, in clipping files, Kansas City Public Library.

the 1920s, when new groups of managers took control of gas and electric operations, upgraded service, and began to supply mass markets at low rates, were gas and electric technologies, public policy, and rapid growth in Kansas City brought into alignment.

After 1900, political stability and financial success among operators of gas and electric firms in Denver and other cities rested on a broad and expanding base of popular enthusiasm for gas and electric service. Sale of gas and electric service, however, depended on more than simple enthusiasm. During the period 1890 to about 1920 or so, thousands of persons in Kansas City and Denver and in every city in North America carried (or better yet sold) information about gas and electric equipment and its applications to residents of their cities. Because purveyors of technical information had backgrounds in such varied activities as architecture, teaching, cooking, real estate, or manufacturing processes, the information dispensed to householders tended to take an exact form. For instance, school officials advised one another on such matters as gas and electricity in classroom ventilation, and home economists advised women on gas and electricity in meal preparation.

Historians of technology apply the concept of technological diffusion to this process whereby increasing numbers of persons become informed about a technical device or method. Improved understanding and more frequent use of gas and electric machines and appliances were part of that diffusion process. Modern scholars label those who provided instruction in the use of machines and appliances the agents of diffusion. No group was more important as agents of diffusion than the sales personnel at the nation's gas and electric companies, particularly those who worked for the innovative Henry L. Doherty. But contemporaries preferred to call those who provided guidance in esoteric matters experts; and the concrete results of their work, observed routinely in the cleanliness, comfort, and convenience of modern schools, plants, and households furnished with gas and electric equipment, was only additional evidence of technological progress. They called it science.

CHAPTER THREE

Agents of Diffusion: Salespersons and Home Service Representatives, 1900–1920

> I want every man to leave this office every morning so interested, so enthusiastic and so ambitious that there will be no limit to your efforts and the results which you can secure.
>
> —Clare N. Stannard
> Denver Gas & Electric Company, June 1906

DURING THE FINAL DECADES of the nineteenth century, residents of Kansas City and Denver perceived gas and electricity as products that were expensive and novel. Electricity for lighting and gas for lighting, heating, and cooking remained the products of magic and mysterious science. Beginning in 1901, however, Henry L. Doherty trained a large sales force and assigned its members the task of boosting consumption of gas and electricity through the sale of stoves, irons, and other appliances to women. By about 1905, members of Doherty's sales force emerged as the principal agents of gas and electric diffusion in Denver. They did so by converting gas and electricity from the realm of science, which was for the elite, into a technology that was part of the ordinary urban and domestic culture and thus accessible to thousands.[1]

1. Historians have published a large and valuable body of literature demonstrating that technology is not science applied. See, for example, Edwin T. Layton, "Mirror-Image Twins: The Communities of Science and Technology in 19th-Century America," *Technology and Culture* 12 (October 1971), 562–580. Contemporaries, however, described gas and electricity as the products of science, the convention that is followed here. For

In 1900, these agents of diffusion were not much better informed than their potential customers about the application of gas and electricity in domestic and business settings. The first order of business, then, is to begin to unravel the processes by which employees at gas and electric firms learned about these exciting new appliances. The second order is to comprehend the methods and concepts that these agents employed as they made their daily rounds of householders, saloon-keepers, retailers, and manufacturers.

Up to 1900, diffusion of gas and electric lights and other appliances had been a slow process. In 1882, a local booster calculated that 350 gas lamps were lighting Kansas City's streets; by 1896, more than 2,700 gas lamps were in service for part of the night, with 209 arc lamps burning all night. During the late 1890s, then, most of the residents of Kansas City and Denver residing outside the central business district walked dark streets during evening hours. Street-paving and establishment of water and sewer service took precedence in urban politics.[2]

---

idiomatic science as a meeting ground, see Mark H. Rose, "Science as an Idiom in the Domain of Technology," *Science and Technology Studies* 5 (Spring 1987), 3–11.

Ruth Schwartz Cowan reminds us of some trends in research on the production and consumption of artifacts. We know a great deal about the work of inventors and system builders, such as Thomas A. Edison and Samuel Insull. Yet we have only a limited understanding of the processes through which light and heat in the form of gas and electricity entered the homes and offices of ordinary Americans. In turn, Cowan recommends that historians of technology focus on what she labels "the consumption junction," the point at which "technological diffusion occurs." Especially important in Cowan's strategy would be specification of the factors that governed the consumption choice—as, for instance, the purchase of a stove by an army officer for a home or for the Ordnance Department. See Ruth Schwartz Cowan, "The Consumption Junction: A Proposal for Research Strategies in the Sociology of Technology," in Wiebe E. Bijker, Thomas P. Hughes, and Trevor J. Pinch, eds., *The Social Construction of Technological Systems* (Cambridge: MIT Press, 1987), pp. 261–280, including quotations on p. 263. Cowan's perceptive strategy is capable of guiding us only so far. Before 1900, reliable and modestly priced gas and electric appliances were not available and most urban residents knew little about their operation. As a practical matter, the consumption junction lacked a site, while the terms of trade and the appropriate behaviors of buyers and sellers lacked definition.

2. *The Leading Industries of Kansas City: A Review of the Manufacturing, Mercantile, and General Business Interests of the "Gate City of the West" to Which Is Added a Historical Sketch of Its Rise and Progress* (Kansas City, Mo.: Reed & Company, 1882), p. 7; *Eighth Annual Report of the Board of Public Works of Kansas City, Missouri, for the Fiscal Year 1896* (Kansas City, Mo.: H. S. Millett Publishing Company, 1897), p. 11. For assessments of the politics of street paving and other municipal projects, see Stanley K. Schultz, *Constructing Urban Culture: American Cities and City Planning, 1800–1920* (Philadelphia: Temple University Press, 1989), pp. 153–205; and Jon C. Teaford, *The Unheralded*

Before 1900, gas and electricity for indoor lighting were available only on a limited basis. Builders of offices, apartments, and houses in the urban west limited gas and electric installations to projects designated for the most selective buyers and renters. Beginning around 1870, developers of fashionable residential districts in the Chicago area included gas hookups along with water and sewer service and paved streets as part of a package aimed at attracting affluent and discriminating buyers. By the 1880s, installation of technical novelties was prerequisite in most cities to securing well-to-do tenants, for whom the high price of novelty was also an investment in the appearance of luxury. In 1880, Horace A. W. Tabor, a flamboyant Denver millionaire, installed gas fixtures in his new, six-story building, which a local observer described as "a grand ornament to the city."[3]

But if novelty and luxury mattered less, or could not be afforded, new buildings did not include gas and electricity. In Kansas City, a warehouse scheduled for construction during 1887 promised only "the most modern ventilating devices." Likewise, a large apartment building, also scheduled for construction in 1887, included "the most recent improvements both as to plumbing and sanitary arrangements" along with "two light wells in the rear of the building."[4]

During the early 1880s, many wealthy householders did not judge gas, for lighting, a necessity. In the absence of a builder or owner who would finance installation expenses, householders, even among the well-to-do, were slow to purchase gas. During 1884, a number of residents of Kansas City constructed substantial homes, and yet many were located beyond city boundaries and beyond gas service. Still others constructed homes within the city but apparently did not install gas lines.[5] Gas was a luxury.

By the late 1880s, following a reduction in prices, wealthier house-

---

Triumph: City Government in America, 1870–1900 (Baltimore: Johns Hopkins University Press, 1984), pp. 227–231.

3. Ann Durkin Keating, *Building Chicago: Suburban Developers and the Creation of a Divided Metropolis* (Columbus: Ohio State University Press, 1988), pp. 71–78; "Tabor Block," in Horace Austin Warner Tabor Papers, FF 474; W. J. Edlebrook to H.A.W. Tabor, March 20, 1880, in Records of the First National Bank of Denver, FF 849, both in SHSC; W. B. Vickers, *History of the City of Denver, Arapahoe County, and Colorado* (Chicago: O. L. Baskin & Company, 1880), p. 240. Concepts of novelty and luxury are based on Mark J. Bouman's unpublished manuscript entitled "City Lights and City Life: A Study of Technology and Urbanity."

4. *Kansas City Star,* May 9, 1887, in Asa Beebe Cross Papers at the University of Missouri at Kansas City (cited hereafter as Cross Papers).

5. *Kansas City Times,* December 31, 1884, and *Kansas City Journal,* January 1, 1885, both in Cross Papers.

holders began to include gas as a routine item in calculations of domestic expenses. Between January 1887 and August 1889, Samuel J. McCulloch, an attorney and real-estate investor in Kansas City, resided in a rented house; he never ordered gas service. Kerosene priced at 80 cents for 5 gallons lit his house. Not until mid-September 1889, following a move to another house, did McCulloch place a deposit with the gas company. He paid $4.80 for gas during the dark month of December 1889, but reduced consumption to $2.24 for May. In McCulloch's household, gas had become a regular expense, requiring entry in his account book along with such items as cable-car fares, a newspaper subscription, and coal for his stove and furnace.[6]

Electricity, however, remained a luxury that McCulloch and most Kansas City residents did not purchase in 1890. Like gas, electricity was sufficiently novel in household budgets and calculations to merit special mention in classified advertisements for apartments and houses. A review of 112 advertisements placed in the Kansas City newspapers for January 15, 1890—roughly halfway through the gloomy and usually chilly winter of the mid-continent region—showed only 29 mentioning gas, steam heat, a range, or electric bells or lights. Although Edison incandescent bulbs were becoming available nationwide, few householders installed electric service. Thus, in 1896, a wealthy householder such as Harry C. James, an executive with the Omaha & Grant Smelting Company in Denver, purchased gas and coal but not electricity for his rented house. "There can be little doubt," a popular writer had predicted in 1890, "that when experience shall have given confidence in . . . [electricity's] trustworthiness, while time shall have rendered its many excellences familiar, it will be adopted by all households."[7]

During the 1890s, electricity, as an indoor service for lighting and other purposes, remained a novelty and an uncertain luxury. Electric service was virtually limited to homes of the well-to-do, to office buildings charging the highest rents, and to hotels catering to fashionable visitors. In 1892, the new and fabulous Brown Palace Hotel in Denver included electric lights and bells in every room, and electric fans drew exhaust air up fifteen shafts to guest rooms for ventilation. Corridors and a court leading to guest rooms, according to the hotel's prospec-

6. Samuel J. McCulloch Records, Folders 1–3, University of Missouri at Kansas City.

7. *Kansas City Star,* January 15, 1890; *Kansas City Journal,* January 15, 1890 (both on microfilm); Harry C. James Papers, SHSC; A. E. Kennelly, "Electricity in the Household," in Kennelly et al., eds., *Electricity in Daily Life: A Popular Account of the Applications of Electricity to Every Day Uses* (New York: Charles Scribner's Sons, 1890), p. 245. (W. Bernard Carlson brought Kennelly's article to my attention.)

tus, were to be "rendered very effective at night by a profusion of electric lights." In 1891, builders of the new Equitable Building in Denver advertised a capacity of 2,060 lights that were powered by their own "dynamo," a popular term for electrical generator. Perhaps because operators of the Equitable Building were themselves concerned about the "novel features in the wiring distribution," the prospectus offered a "description of the wiring . . . from the dynamo to the lamp." In older but still-exclusive buildings, such as the Ernst and Crammer Building, which housed bank offices, attorneys, and insurance agents, operators reported that both gas and electric service were available and that the local supplier of equipment had "furnished most of the large buildings erected in Denver during the past three years."[8] Although electricity was judged unpredictable, during the 1890s electric lighting moved into the realm of an expected amenity among members of higher-income groups.

After 1890, electric lighting began to enter public buildings. Because electricity was uncertain and mysterious, architects hedged their specifications in several ways. The Jackson County Courthouse, completed in 1892, featured 1,000 incandescent bulbs. But the architect for the courthouse (which was located in downtown Kansas City) followed a practice common in the design of private buildings by having gas lighting installed alongside the electric lights. Designers of a new high school in Kansas City installed lighting fixtures that combined electric and gas service, risking fire—as eventually occurred in October 1894. By 1898, electricity was the exclusive form of illuminant for the new public library in Kansas City, but it was delivered in fixtures made to appear identical to old-fashioned gas lamps. The future in this case was packaged as the past—and candles, set in chandeliers around the mantel of the rotunda, guaranteed the presence of the past on formal occasions, as well as a modicum of light in the event of another equipment failure.[9]

8. *The Brown Palace Hotel: Denver, Colorado* (n.p., c. 1892), in Claude K. Boettcher Papers; Brown Palace Hotel Company, Denver, *Prospectus* (Denver, 1892)—both in SHSC; *Descriptive and Illustrative of the Equitable Building, Denver, an Acknowledged Type of American Architecture, and Probably the Foremost Example of Modern Fire-Proof Construction* (New York: Exhibit Publishing Company, 1892); J[ackson] D. Dillenback, *A Souvenir: The Ernst and Crammer Building and Its Tenants* (Denver, Colo.: W. F. Robinson & Company, 1891), p. 9. For similar patterns in Chicago, see Platt, *Electric City,* pp. 22–23, 34–36.

9. "Specifications for Steam Heating, Power Plant, Engine, Dynamo, Elevators to Be Erected for the New County Court House Situated Between 5th and Missouri Ave., and Oak and Locust Sts. at Kansas City, Missouri"; "Schedule of Combination Gas and Electric Fixtures Required to Finish the Courthouse of Jackson County, Mo."—both in

Fig. 2. Brown Palace Hotel, Denver, mid 1890s. Hotels catering to higher-income travelers installed electric lights as part of their effort to appear up-to-date. (Courtesy, Denver Public Library, Western History Department)

Rapid installation of gas and electric service in public buildings such as schools, libraries, and courthouses appears an anomaly. Urban leaders had determined that novelty and luxury were to enter the domains of frugality and restraint. One suspects, however, that city and county governments acted promptly to install electric lights as part of a commitment in every city to boost local industry and to appear up-to-date in the race for urban greatness. Increasingly, business and professional leaders in American cities looked to science as an agency capable of bestowing a portion of that prosperity and re-

Cross Papers; *Kansas City Star,* October 23, 1894 (microfilm); *The New Kansas City Public Library, 1898* (Kansas City, Mo.: Architecture & Builder Publishing Company, 1898), an unpaginated collection of photographs.

Fig. 3. Denver residence. Before 1900, electric lamps provided only modest levels of illumination at high prices. Electric lighting was both a novelty and a method for wealthier householders to signal their commitment to a process that contemporaries labeled progress. (Courtesy, Denver Public Library, Western History Department)

nown. Perhaps in a similar fashion, the popular association of gas and electricity with science facilitated diffusion of modern electric lighting to libraries and schools, where, many believed, the mysteries of science would be clarified and turned into useful products for the next generation.

During the 1890s, electricity emerged as a topic in the rhetoric of civic boosters. As early as 1892, J. H. Ernst Waters, an engineer with experience in applying electricity to mining, constructed a four-story building in Denver and equipped it with electrical service for small manufacturers, such as brass finishers and lithographers. "We cannot but believe," ran a report in an engineering journal, "that such facilities must draw to Denver from the East and elsewhere many promising young industries in search of power and a market." Subsequent reports of electrical applications in manufacturing were equally enthusiastic. "Coal no longer King," ran one story in 1896, and "Electricity the Craze" ran another that same year, emphasizing further the evidence of "science in greatest perfection" and that "development in Colorado is greater than elsewhere."[10]

For all of the hoopla that surrounded gas and electric service, as late as 1900 such services remained outside of the domestic and work experiences of a majority of urban residents. Consumers of gas and electricity tended to be among that small group who occupied prestigious office buildings and whose lovely homes and apartments were located in neighborhoods established for the well-to-do. In February 1897, an editorial writer in Kansas City observed that, among the small number of subscribers to the new gas company, "nearly all belong to the class of established householders." Most residents of Kansas City and Denver, however, were beginning to enjoy the light and heat of gas and electric service as shoppers in stores catering to higher-income residents, as visitors to public buildings and exclusive hotels, as pedestrians passing through the central business district, or perhaps as servants in the homes of better-off families.[11] As an indoor service, gas and electricity appeared to be expensive, complicated, and uncertain—still representative of "the splendid triumph of science."

The popular association of gas and electricity with science reinforced the notion that only members of the elite could participate in these new technologies. Contemporaries spoke of electricity in particular as an invention produced by magic and mysterious science—novel and impenetrable, and thus a product at which even "all scientists gaze in wonder." As an idiom, science was part of a process that historian Charles E. Rosenberg identifies as "an unquestioning faith in the unambiguous virtue of progress." By 1900, few doubted the practical benefits of progress in the form of gas and electric service. The association of gas and electricity with idiomatic science suggested

10. "The Sheridan Electric Power and Office Building, Denver, Col.," *Electrical Engineer* 14 (December 14, 1892), 568–570, an item identified for me by W. Bernard Carlson; *Rocky Mountain News,* February 14, 1896, and March 30, 1896; *Denver Republican,* February 12, 1893—all in Denver Public Library.

11. *Kansas City Journal,* February 15, 1897, in Snyder Collection; William R. Leach, "Transformations in a Culture of Consumption: Women and Department Stores, 1890–1925," *Journal of American History* 71 (September 1984), 324.

luxury, however, which meant that only the well-off would purchase them, at least for the near future.[12]

In Denver, Henry L. Doherty and his associates fashioned a strategy for boosting consumption of gas and electricity. As first introduced in the previous chapter, Doherty assembled a large force of salespersons and instructed them in the merits of his three-part rate and about the uses of gas and electric appliances. Next, he directed them to undertake massive sales campaigns that took account of rapid and extensive urban and industrial change. One result of Doherty's efforts was to shape a large and disciplined organization capable of shortening the period of time between the introduction of an appliance and its rapid diffusion among Denver's householders. Another and related result of Doherty's efforts was to define gas and electric appliances as appropriate for women alone. Fortunately, the company printed the deliberations that took place among sales representatives during the early years of promotional work. That material, in combination with other manuscripts and publications, permits a unique glimpse into the organization and education of an urban sales force.

In 1900, the shortage of persons who were trained and experienced in the design and operation of gas, and especially electrical systems, was nationwide in scope. Between 1882 and 1885, the Massachusetts Institute of Technology and Cornell University had initiated programs in electrical engineering. By the late 1890s, according to historian Robert Rosenberg, "every respectable school—and many seeking respect—had instituted some sort of instruction in the new science." A large number of universities also offered courses of study in civil and mechanical engineering. Those two fields, along with the newer field of electrical engineering, were the areas best suited to prepare young engineers to solve problems in complex areas, such as the design of steam turbines and long-distance, high-voltage transmission. By 1900, universities and colleges had awarded 1,269 degrees in these three areas nationally, but none of the persons holding those degrees was a member of the senior management, sales, or technical staffs at the Denver Gas & Electric Company.[13]

12. As reprinted in the *Kansas City Star,* June 4, 1950; Charles E. Rosenberg, *No Other Gods: On Science and American Social Thought* (Baltimore: Johns Hopkins University Press, 1976), pp. 7, 137. See also David E. Nye, *Electrifying America: Social Meanings of a New Technology* (Cambridge: MIT Press, 1990), pp. 242, 382.

13. Thomas P. Hughes, *Networks of Power: Electrification in Western Society, 1880–1930* (Baltimore: Johns Hopkins University Press, 1983), pp. 143, 144–146; Robert Rosenberg, "The Origins of EE Education: A Matter of Degree," *IEEE Spectrum* 21 (July 1984), 63.

Knowledgeable people were scarce at all levels, and the demand for persons who could maintain the system, selling and adjusting machines and directing the administration of sales and service forces, was even greater. To remedy this lack of knowledge about gas and electric operations among employees at every level, Doherty turned his attention first to Denver's Manual Training High School, the city's technological high school. In 1900, Doherty hired five graduates; and between 1901 and 1905, he hired additional graduates, employing the men as meter readers and the women as home service agents. After 1905, Doherty no longer sought graduates of Manual or the city's other high schools, but instead began to employ those graduating with honors from the nation's engineering colleges. During the period 1906 to 1915, as Doherty expanded operations into a large number of cities, he hired an additional 180 graduates, all male, with degrees in engineering fields including mechanical, electrical, and chemical engineering. The idea behind hiring graduates of technologically oriented schools was to take advantage of their backgrounds in a craft or a specialized area of engineering expertise, as well as their orientation toward problem-solving.[14]

Technical background and personal inclination were only prerequisite to employment at Denver Gas & Electric. Doherty also created a program of instruction and on-the-job training for new employees. During the period up to around 1904, the training program was informal. Doherty assigned the title of "apprentice" to recruits and provided a brief course of instruction before making assignments to such posts as meter reader and home service agent. Doherty also encouraged apprentices to enroll in correspondence courses, financing tuition up to 100 percent for those completing their studies.

14. Roy G. Munroe to Robert I. Marshall, October 28, 1947; William P. Strobhar, comp. and ed., *A Manual of Communities Served by Corporations Operated by the Henry L. Doherty Organization* (n.p.: Doherty Men's Fraternity, 1920), p. 90, both in Munroe Files; National Electric Light Association, Denver Section, *Bulletin* 1 (July 1909), 1–2 (hereafter *NELA Bulletin*). Roy G. Munroe Scrapbook, 1893–1970 (microfilm), Denver Public Library (cited hereafter as Munroe Scrapbook). Schools like the Manual Training High School should not be confused with vocational schools, which also emphasized training in a skill but without requiring coursework in the sciences, mathematics, and other elements of a general high school education. For the demographics, patterns of recruitment, and courses of instruction at a high school in Pittsburgh that appears to have been similar in orientation, see Ileen A. DeVault, *Sons and Daughters of Labor: Class and Clerical Work in Turn-of-the Century Pittsburgh* (Ithaca, N.Y.: Cornell University Press, 1990), pp. 24–47. Finally, see Bruce E. Seely, "Research, Engineering, and Science in American Engineering Colleges: 1900–1960," *Technology and Culture* 34 (April 1993), 344–359, on the emphasis in engineering education before 1945 on solving the practical problems of industry.

Promotions and reassignments came rapidly for men, with one of the members of the Manual class of 1900 reporting within four years to positions as meter reader, bill collector, assistant bookkeeper, and sales representative. A growing number of customers and increased sales created a demand for additional staff, and mangers offered temporary promotions to apprentices exhibiting talent and energy, making the assignment permanent on evidence of satisfactory performance. Women, even graduates of Manual, entered employment at Denver Gas & Electric and in the utility industry generally on a narrower track. Their ability to learn about various dimensions of sales, service, and management was restricted. As home service agents, energetic and talented women could aspire only to posts as supervisors of home service.[15]

Beginning in 1904, Doherty placed a more formal cast on the training program. Recruits were awarded the title of Cadet Engineer (later Junior Engineer) rather than "apprentice," and the course of instruction was prescribed instead of left to chance. Nonetheless, much of the training remained practical in its orientation. Company officials rotated cadets through several departments, such as accounting or sales, during a period of two years. Although cadets were "not hired to do work that is ordinarily paid for by the hour," the expectation, according to a report prepared years later, was that they would "do hard work or dirty work, perhaps extra work with long hours at times, and occasionally under disagreeable conditions."[16]

Formal instruction, in a classroom setting, complemented on-the-job training. Classes convened at company headquarters every Monday and Thursday from 7:30 P.M. to 9:00 P.M. during the two-year cycle. Cadet engineers earned 50 cents for each class attended. Recruits from Manual Training High School who remained in the program during its first years also were invited to attend, but without pay. The course was titled the "Doherty School of Practice," distinguishing it rhetorically from collegiate instruction, which was already judged overly theoretical. At each session, managers lectured on such specialized aspects of gas and electric operations as accounting and of course sales. Lecturers asked specific questions and insisted on specific answers. One participant reported that much of the informa-

15. Munroe to Marshall; Munroe Scrapbook. For the training and ideology of sales organizations in other firms, see Olivier Zunz, *Making America Corporate, 1870–1920* (Chicago: University of Chicago Press, 1990), pp. 179–197.

16. Munroe to Marshall; Strobhar, *Manual of Communities,* p. 90; H. B. Shaw, "The Doherty Training Schools," *Doherty News* 5 (March 1920), 5–6, Munroe Files.

tion was "well above our heads" but that managers expected them to "pick up as much as we could."[17]

No group of employees was subjected to more intense training than members of the sales force. By 1904, Doherty had shaped a regular program of instruction in sales. Every Tuesday evening, sales personnel reconvened at company headquarters for instruction in marketing and other aspects of operations. Doherty's executives provided detailed instruction in the practical dimensions of selling gas and electric products and service, followed by exams that were graded and returned. "Describe the method used in testing gas meters," was a question put to a representative during the evening of June 11, 1907, and "What is the average cost of a furnace connection to a kitchen?" was another. At a point when factory operators were beginning to convert from steam and belts to electric drive, the following exchange on February 12, 1908, was no doubt valuable. Regarding the problem "What is the best proposition for a consumer operating a machine shop with several machines, individual drive for each machine or a single large motor for all machines?" salesmen were advised to resolve the matter according to the "size of the shop," because "if the consumer can stand the investment for individual drive, he will be able to cut down his power bill, as there will be no shafting." Once in a while, Doherty attended meetings, presenting remarks on topics such as the importance of courtesy and the equal importance of "the sales end of our business to make our profits, to make things better and bigger."[18]

Education and exhortation were only the first steps in creating a disciplined sales force and boosting consumption. As early as 1901, Doherty launched a program of sales solicitations built around newspaper advertising, cooking schools, and house-to-house promotions and demonstrations. Newspaper advertising was important as a preliminary step in the sales process, he contended, because "it appeals to your prospective customer . . . and may strike him some day in a receptive mood and he begins to 'get religion,' gives your solicitor an interview and becomes a victim." In the case of gas for industrial purposes, he argued, demand "can only be made universal by creating a universal want for it, and all the people cannot be reached except by fetching 'ads,' so placed and designed that they will be read." Doherty did not believe, however, that advertising sold prod-

17. Munroe to Marshall; Munroe Scrapbook.

18. Munroe Scrapbook; *DG&E Bulletin* 2 (June 11, 1907); *DG&E Bulletin* 3 (February 12, 1908); Minutes, Meeting of Office Association, September 7, 1904; Commercial Council Meeting, July 24, 1906—all in Munroe Files.

ucts. "I never heard of anyone coming into the office and saying they wanted a gas stove because they saw an advertisement in the paper," he wrote fellow executives.[19]

With local newspapers carrying his advertisements, Doherty insisted that sales personnel make demonstrations that would actually close deals. Demonstrations of the appliance, especially house-to-house demonstrations, appealed to Doherty "more strongly than anything else." One method that attracted attention, Doherty contended, was for a salesman to "borrow a kitchen equipped with a gas stove, [and] invite in all the neighbors who have no stoves." Another method of gaining favorable attention was to offer demonstrations in the city's churches. "We furnish the demonstrator," he reported, as well as "the range and the food to any church in town which wishes to give a 'cooking demonstration.'"[20]

An increase in sales did not necessarily follow each new series of advertisements in newspapers and demonstrations in homes and churches. High prices and lack of familiarity with new appliances stood in the way. One salesman recalled that many households were not equipped for gas and electric service. Still other householders, though connected, were "afraid they would be electrocuted and both electricity and gas were expensive." Only during the summertime, when the heat of a coal stove became oppressive, would valves for the small gas-fired burners on top of coal stoves be turned on. But the salesman also remembered that "women did as little cooking as they could with gas because of the high price and fear of it." During the next few years, sales personnel discovered that appliances such as electric irons and automatic gas water heaters encountered a similar level of resistance brought about by a combination of fear and lack of knowledge regarding cost and methods of operation.[21]

Early on, however, Doherty had determined not to allow the sale of gas and electric appliances to depend on the fears of local residents or the idiosyncrasies of sales personnel. Instead, he instituted a mix of regulations, checking procedures, and cash incentives aimed at developing a courteous and disciplined work force. Doherty required representatives to appear at headquarters each morning by 8:15 for roll call, which was followed by a sales meeting. Sales representatives

19. "Interview with Roy G. Munroe," c. 1970s, Munroe Files; Glenn Marston, comp., *Principles and Ideas for Doherty Men: Papers, Addresses, and Letters, by Henry L. Doherty* (New York: Henry L. Doherty & Company, 1923), 154–155.

20. Marston, *Principles and Ideas,* p. 205.

21. Munroe Scrapbook. Also, Roy G. Munroe to George E. Lewis, August 18, 1950; Roy G. Munroe to Elmer W. Cone, April 15, 1949—both in Munroe Files.

were expected to leave the office by 9:15; office boys reported the names of laggards to an executive. Six days a week, representatives made sales calls—Doherty's preferred method—from around 9:15 A.M., until the lunch hour, and then from 1:15 to 5:15 P.M. Reports listing contacts and results were filed on a daily basis. Inspectors from headquarters in New York, whom representatives named "spotters," checked once or twice a year that reports and the actual number of calls matched one another. Rules regarding behavior during a sales presentation were specific—for instance, at no time was a representative to smoke cigarettes in public view. Still another regulation required representatives to remove rubbers from their shoes on rainy and snowy days when householders opened their doors. Doherty wanted to assure prospective customers that carpets and floors would remain clean, and also increase chances that representatives would be invited to step inside.[22]

Nor did sales personnel enjoy much discretion regarding the products to be emphasized or the terms of a contract. Occasionally, executives offered directives in a general form. Representatives were told at the meeting held on June 28, 1906: "I want you gentlemen in your house-to-house work to make sure and mention the various ways in which gas and electricity can be applied in addition to the present equipment." On other occasions, executives provided specific instructions about prices and products. On June 25, representatives learned that "we will specialize this week on industrial fuel and interior illumination." In addition, the manager informed them on June 28: "I want every representative to specialize on gas range sales on the residence territory, and make a report of your work on July 7." At still other meetings, managers presented representatives with a list of "Don'ts" to keep in mind during sales contacts, such as, for instance, that the company "will not assume any responsibility as far as sign construction is concerned" and "will not replace lamps in a sign which requires a balloon or an airship to reach it."[23]

Doherty favored cash prizes as a reward for exemplary efforts and as another method for keeping the attention of sales personnel focused on boosting consumption of gas and electricity. For instance, on June 9, 1906, an executive awarded three checks in amounts between $5 and $15 to salespersons "for having the greatest number of points from the sale of gas ranges during this last month." On June 28, Doherty announced a prize of $50 for first place, $30 for second,

22. Undated interview with Munroe, Records of the PSCC.
23. *DG&E Bulletin* 1 (June 1906), 12–14; *NELA Bulletin* 1 (July 1909), 3.

and $20 for third to be awarded to the representatives who sold the "greatest percentage of the total number of range prospects turned in a few days ago." Successful representatives often were promoted into management positions, usually in sales, with other companies in the group. Each promotion served as another opportunity, similar to the award ceremonies, to refocus attention on building demand for gas and electricity. On June 20, 1906, for example, representatives learned that "Mr. Mateer has received a well merited promotion, and leaves the last of this week going first to Shawnee, and later to Montgomery." The executive added: "[It] is greatly appreciated . . . to have the company reward good work and splendid effort."[24]

Beginning in mid-1906, executives of the Denver Gas & Electric Company added another feature to the load-building efforts, designating three representatives as "Supervisors of Service." By 1908, seven supervisors were at work. This title, as distinct from such customary forms as "representative," "salesman," or the outmoded title "solicitor," appeared desirable "in order to make an easy entre to the customers premises [*sic*]." Supervisors did not handle routine new business, but were charged instead "to work upon the accounts of old consumers." Each supervisor focused on 300 accounts during a four-month period, thus reopening sales opportunities with up to 900 customers in the course of a year. Where, for example, a householder was cooking both on a small gas range and on a coal range, the supervisor aimed to sell a new gas range and a water heater. The overriding concept in this approach was to persuade a customer who "had been in the habit of spending $1.00 per month for gas and $4.00 per month for coal . . . to spend the entire $5.00 per month for Fuel Gas." Because these customers were already attached to the company's lines, an executive reported that increases in consumption and revenue were "accomplished without expense on our part for construction." Sales commissions still had to be paid, yet the net result, according to the calculations of the same executive, was "about $2.00 in revenue for each $1.00 spent in solicitation."[25]

Between 1900 and 1915, the sales force at the Denver Gas & Electric Company emerged as the major agency in the city for disseminating knowledge about gas and electric service. In great measure, its task

24. *DG&E Bulletin* 1 (June 1906), 7, 11, 13.

25. Minutes, Doherty Operating Company, 60 Wall Street (February–October 1908), pp. 150–151, 207–208, in Records of the PSCC; "Service Supervisor Work in the Denver Gas and Electric Company," *Electrical World* 50 (November 2, 1907), 881.

consisted of asserting the superiority of centrally distributed gas and electricity over the burning of coal and petroleum distillates—fuels that were cheap, plentiful, and well understood—in kitchens, offices, workshops, and basements. At first, possessing no special training for assessing the preferences of customers, sales personnel turned naturally to the evolving standards of professional and business leaders in their city. By 1900 or so, as historians Stanley K. Schultz, Joel A. Tarr, and Nancy Tomes demonstrate, experts in law, engineering, landscape architecture, and medicine had created institutions capable of shaping and controlling portions of the environment, especially pertaining to water and sewer service and the elimination of waste and dirt from streets, gutters, households, and one's own body. In turn, educated and wealthier residents were beginning to assume that it was possible to shape their own environments, creating commercial and private buildings and a city as a whole that guaranteed personal cleanliness, comfort, and convenience.[26] Members of Doherty's sales force transformed these general desires, along with subtle invocations of gender, into the purchase of specific appliances and machines such as irons, fans, stoves, and light-bulbs.

The rhetoric and logic of sales presentations emerged from the day-to-day experiences of Doherty's representatives. As sales personnel familiarized themselves with different sections of the city, they began to correlate tastes and buying habits. Representatives spoke routinely of the purchases made by businessmen located downtown and of the special needs of druggists, saloon-keepers, and restaurant operators. During the course of reports made orally each day on sales contacts, representatives learned about the heating and lighting needs

26. Joel A. Tarr, "Sewerage and the Development of the Networked City in the United States, 1850–1930," in Joel A. Tarr and Gabriel DuPuy, eds., *Technology and the Rise of the Networked City in Europe and America* (Philadelphia: Temple University Press, 1988), pp. 159–185; Nancy Tomes, "The Private Side of Public Health: Sanitary Science, Domestic Hygiene, and the Germ Theory, 1870–1900," *Bulletin of the History of Medicine* 64 (Winter 1990), 509–539; Schultz, *Constructing Urban Culture.* For additional examples of the relationships among politics, technological systems, and the desire and perceived ability to shape a healthful built environment, see John C. Burnham, "Essay," in John D. Buenker, John C. Burnham, and Robert M. Crunden, eds., *Progressivism* (Cambridge, Mass.: Schenkman Publishing Company, 1977); Martin V. Melosi, *Garbage in the Cities: Refuse, Reform, and the Environment, 1880–1980* (College Station: Texas A&M University Press, 1981); and Marilyn Thornton Williams, *Washing "The Great Unwashed": Public Baths in Urban America, 1840–1920* (Columbus: Ohio State University Press, 1991). Finally, see Richard L. Bushman and Claudia L. Bushman, "The Early History of Cleanliness in America," *Journal of American History* 74 (March 1988), 1217, for the observation that "by the end of the century, . . . cleanliness versus dirt [had] entered deeply into middle-class people's judgements of other human beings."

of Jewish merchants on Larimer Street and about the conversion to higher-priced ranges among Chinese households, which an executive interpreted as evidence that "these people are progressive." Nor could a representative fail to become aware that the sale of appliances to create all-gas kitchens was taking place most rapidly among households located in new homes in the outlying districts.[27]

After 1906, knowledge of urban change developed by Doherty's sales representatives began to play an explicit role in explaining (and planning) for the pace of gas and electric sales in different neighborhoods. By late 1908, sales representatives had identified a relationship between low income in several districts and continuing reliance on coal for cooking. Representatives had also reached the conclusion that "gas lighting is confined almost altogether to the older houses and stores in the poorer sections of the city." Beginning in June 1909, moreover, officials at Denver Gas & Electric conducted surveys of the number of persons crossing busy intersections in the central business district, using the data collected to speak with retailers about the advantages of store lighting and electric signs. The city's spatial order and daily population movements thus emerged as analytic tools in the process of boosting loads on the company's gas and electric systems.[28] The ability to translate simple empiricism into suspected correlations between urban change and the sale of gas and electric appliances had thus emerged as another dimension of Doherty's training programs.

By about 1905, managers and representatives had articulated some standard ideas aimed at promoting the sale of appliances and boosting the consumption of gas. Salesmanship was assuming a routine form, one guided by notions of cleanliness, environmental control, convenience, and gender. At the daily sales meeting held on June 29,

27. Impressions gained from a review of volumes 1–3 (1906–1908) of the *DG&E Bulletin.* Remarks of sales personnel were published in a monthly (later weekly) bulletin. Publication of observations made at daily meetings created an opportunity for each salesperson to review successful sales strategies. Of equal value for those who had failed to comprehend the matter during oral reports was the ability to restudy the stated relationships between gas and electric operations and the preferences of particular groups, such as merchants and householders, who were spreading rapidly across the urban landscape. Historically minded scholars have not paid much attention to the diffusion process, including crucial agents of diffusion such as sales personnel. For the modest state-of-the-art of the literature in this area, see Patrick Kelley and Melvin Kranzberg, eds., *Technological Innovation: A Critical Review of Current Knowledge* (San Francisco: San Francisco Press, 1978), pp. 119–150; and Lawrence A. Brown, *Innovation Diffusion: A New Perspective* (New York: Methuen, 1981), pp. 176–196.

28. Roy G. Munroe, "N.C.G.A. Course III," Munroe Files; *DG&E Bulletin* 3 (November 30, 1908), 1; NELA, Denver Section, *Bulletin* 1 (July 1909), 4.

1906, for example, an executive asked for a reiteration of arguments in favor of gas for cooking. Economy of effort comprised part of the promotional routine, particularly for busy housewives. "I often tell the customer that we are going about doing good," observed a sales representative, "because we cut down the housewife's work and labor as much as 50 percent." Similar in content was the report of a representative who pointed out: "Where a lady has a day's work to do, the gas range is good for an hour's saving of time."[29]

Still another argument for gas appliances was cleanliness. A gas stove eliminated the smoke, dust, and heat of a coal stove. Two sales representatives employed identical language, informing colleagues and management that they spoke of "the cleanliness of the gas range." Along those lines, then, it was possible to combine considerations of economy with an emphasis on cleanliness. "Usually a housewife will take one day of the week for cleaning the woodwork of the kitchen," contended a representative, but "where she uses gas her kitchen is not nearly as dirty at the end of two weeks as it is when using coal at the end of one week."[30]

Gender was also an explicit factor in promoting gas and electric consumption. Gender worked on two levels. The first and most obvious was the creation of an indirect linkage between conventional expectations about domestic responsibilities and the purchase of a new gas stove or other appliance. So subtle were matters at this level that representatives often lagged behind consumers. "I talked with a lady yesterday," one representative pointed out, "whose husband did not wish a gas range." We tried to convince him of its advantages, he added, and yet the man remained unpersuaded. According to the sales representative, the husband asked his wife "what she thought about it." She responded, again as reported by the sales representa-

29. *DG&E Bulletin* 1 (May 1906), 14. See also Ruth Schwartz Cowan, *More Work for Mother: The Ironies of Household Technology* . . . (New York: Basic Books, 1983), pp. 98–101. For the argument that gas and electric appliances reinforced the household division of labor according to gender, see Charles A. Thrall, "The Conservative Use of Modern Household Technology," *Technology and Culture* 23 (April 1982), 175–194. For the related and more general observation that "new electrical inventions and ways of thinking about electricity were given shape and meaning by being grafted onto existing rules and expectations about the structure of social relations," see Carolyn Marvin, *When Old Technologies Were New: Thinking About Electric Communication in the Late Nineteenth Century* (New York: Oxford University Press, 1988), pp. 233–234. Carolyn Goldstein's case study, "Mediating Consumption: Home Economics and American Consumers, 1900–1940" (Ph.D. dissertation-in-progress, University of Delaware), a portion of which she generously shared with me, makes several of these points explicit.

30. *DG&E Bulletin* 1 (May 1906), 14.

Fig. 4. Denver Gas & Electric Company employees, February 1908. Although employees were often college-trained, managers still required them to attend evening classes at company headquarters. Sales were a favorite topic, with employees expected to take notes and pass exams. Regardless of academic background or test scores, however, job assignments were made according to gender, with men occupying managerial, technical, sales, or labor positions, and women taking posts as home service agents and clerical workers. (Courtesy, Public Service Company of Colorado)

tive, "that she had but two burners to use, and while she was cooking part of the meal, the rest was getting cold."[31] Messages of comfort and cleanliness and of the different responsibilities of men and women toward one another must have been lodged routinely in the ordinary bargaining of husbands and wives.

Gender operated on another and more explicit level. Typically, a man, whether a representative or a service supervisor, made calls at the homes of women who were housewives, advising on the advantages of gas and electric appliances; once in a while, salesmen even prepared meals in the homes of sales prospects. But Doherty employed women to follow up in a more systematic fashion on those sales calls. In practice, a "home service agent"—the title assigned to

31. Ibid.

Fig. 5. Denver, looking up Sixteenth Street toward the State Capitol, 1911. (Courtesy, Public Service Company of Colorado)

women employed as members of the sales force—worked a territory in conjunction with a representative and a service supervisor. A salesman directed his female counterpart to a particular household in order to close a sale or to demonstrate new and improved methods of using installed appliances. By 1907, as part of an effort to acquaint householders with the type of work that home service agents would

perform, the company published a brochure announcing that each had "taken a special course in domestic science so that she may tell you how to do more with your gas range than you can do with any other cooking appliance—and do it better." In November 1908, then, a service supervisor reported that two consumers were "on my list of low consumption, and upon investigation I found out it was because they could not operate the range satisfactorily." He determined to have "Miss Lull go out . . . and give them a demonstration, and from Miss Lull's report I believe that this will materially increase their consumption."[32] The conventions of gender relationships were thus embedded in the demonstration and sales processes of Doherty's sales efforts.

Within a few years of creating and educating a large sales staff, arguments regarding comfort, convenience, or economy came together easily with the conventions of gender. Promotion of electric irons created one of many outlets for this combination of themes. In 1905, Doherty and his executives decided to promote electric flatirons, which consumed 600 to 700 watts per hour. (By comparison, a standard light bulb consumed a mere 64 watts.) According to the recollections of one representative, members of the sales force were "skeptical concerning the possibility of inducing women to use such appliances for ironing." Nonetheless, this representative demonstrated an iron to anyone who allowed him to do so, checking on fuses (which frequently blew out under such a high wattage) and then leaving the iron for a home trial. Initially, he found that housewives feared the iron, expressing concern about shocks and scorched linens. Home trials lasted a couple of days but failed to win converts and sales, for upon his return housewives would "order . . . me to remove the apparatus, saying they would positively have nothing to do with it."[33] Not for six weeks did this salesperson secure a permanent installation.

By early 1907, however, Doherty's executives had developed techniques to distribute large numbers of electric irons on a regular basis. In brief, they paved the way for their sales force by advertising the advantages of the electric iron in terms of cleanliness, comfort, and convenience. Then they ordered the force of salesmen and home service agents to make countless house calls and demonstrations. In 1907, for example, literature prepared by the company included the observation that "the day when the clothes are washed has been one of the most dreaded of the week." A new electric iron, according to the company, would "remove that feeling which tangles nerves and tires bodies." Nor was there any expense in operating an electric iron, as this notice ran, "because the time saved will more than pay for the current used." Between 1907 and 1909, service supervisors handled promotion of electric irons, offering them for sale at a price of $5 apiece to households already connected to company lines. By Novem-

32. Munroe Scrapbook; *DG&E Bulletin* 3 (November 30, 1908), 6.
33. Roy G. Munroe to J. F. Bynon, June 23, 1925, PSCC.

ber 1909, the company had sold 9,000 irons, and an executive estimated that approximately 50 percent of Denver households owned an iron.[34]

In 1915, Doherty's executives launched their last campaign to promote the sale of electric irons. The General Electric company was a co-sponsor—an increasingly common feature in promotional efforts. Each morning, representatives loaded horse-drawn wagons with irons and traveled up and down city streets ringing doorbells and awaiting the women who answered. According to one account, representatives "explain[ed] that this was a free one-week trial . . . and that we would be back within a day or two to show her how to operate it." Representatives returned on bicycles, in which case "it was obvious that we could not transport the flat iron back to the office." Still, many refused to make a purchase, and irons were left for lengthier home trials, which again did not always culminate in a sale. During this campaign, which lasted three months, a representative managed to sell 277 irons and was awarded a trip to San Francisco and the meeting of the National Electric Light Association for himself and his family.[35]

By 1915, the sales force at Denver's combined gas and electric company had achieved a considerable degree of sophistication in planning and implementing promotional activities. For example, between November 29 and December 4, 1915, executives sponsored Electrical Prosperity Week. Similar to the electric flat-iron campaign co-sponsored with General Electric, Electrical Prosperity Week was launched in coordination with the Society for Electrical Development, which was a coalition of manufacturers and utility operators located around the nation who were seeking to boost the sale of appliances and electric current. In Denver, Doherty's employees managed a large booth filled with appliances at the Electric Show and Home Industrial Exposition held in the City Auditorium. All the while, remaining sales personnel walked their rounds, and a former domestic science teacher continued to make daily presentations at company headquarters on "every phase of scientific housekeeping." Notice of these demonstrations included coupons allowing a price reduction on appliances, particularly on an electric vacuum cleaner priced at $50 and payable monthly. Representatives sold 801 appliances during the

34. Munroe Scrapbook; "Service Supervisor Work in the Denver Gas and Electric Company," *Electrical World* 50 (November 2, 1907), 881; "Progress with Heating Appliances at Denver," *Electrical World* 52 (October 31, 1908), 950; "Heating Devices and Increase of Residence Revenue in Denver," *Electrical World* 54 (November 18, 1909), 1239.

35. Munroe Scrapbook.

Fig. 6. "Cook with Gas" vehicle, c. 1910. Sales executives were usually alert to new opportunities to promote consumption. (Courtesy, Public Service Company of Colorado)

week, an increase of 538 over the same period the year before. In December 1915, a company publicist announced that Electrical Prosperity Week had been "an all-around success."[36]

Between 1901 and 1915, then, executives of the Denver Gas & Electric Company had organized, educated, and disciplined a large and increasingly specialized sales force. At first, neither managers nor representatives enjoyed specific training or experience in the selection of appealing notions for sales promotions. Instead, they brought a set of cultural assumptions that had been encoded in their memories and were carried without critical examination to each sales presentation. Ordinary representatives, along with service supervisors and home

36. [*New Business Bulletin*], December 20, 1915, 5–9; Munroe Files: "Denver Company Gives Housekeeping Demonstrations," *Electrical Review and Western Electrician* 62 (May 24, 1913), 1050.

service agents, spoke with women in terms of cleanliness, comfort, convenience, and economy of operation.

Salespersons never allowed their customers' decisions to revolve solely around these general propositions. The creative dimension of promotional activities rested on the ability of sales personnel to link general notions such as comfort and cleanliness with the personal concerns of each householder in particular. Marketing electric irons with the reminder that "the day when the clothes are washed has been one of the most dreaded of the week" was one way to connect the general and the particular. Still another was the contention, made before 1910, that an auxiliary gas furnace would eliminate cold rooms if "the [coal] furnace is kicking up," and thus reduce the "risk of catching a bad cold." The idea itself along with its colloquial phrasing spoke to a general interest in comfort and cleanliness in a manner that most could no doubt appreciate. One suspects, too, that the group most sensitive to messages about the health benefits of gas heaters were the affluent householders who were coming to believe that installation of gas and electric appliances would allow them to shape a healthful environment in the middle of an industrializing city.[37] By about 1915, moreover, salespersons at the Denver Gas & Electric Company had launched a process of demystifying gas and electricity from the rank of a science to that of a technology.

After 1900, members of several other groups began to spread the news about exciting gas and electric appliances. In particular, community-builders and educators emerged as prominent agents of diffusion on the urban scene. Necessarily, they spoke to specific constituencies. Community-builders focused on the sale of homes to affluent families seeking a lovely house along the city's periphery. Educators, the subject of the next chapter, sought to prepare students for careers associated with gas and electric machinery. Educators also worked to achieve enhancement of learning, protection of eyesight, and elimination of disease.

37. Denver Gas & Electric Company, "Here's a Way to Stop Furnace Troubles" (advertisement), Munroe Files; Munroe Scrapbook. See also Tomes, "The Private Side of Public Health," 509–539.

CHAPTER FOUR

Preparation of Producers and Consumers: Public Schools, 1900–1930

> The result was the immediate disappearance of all the stuffy condition and bad smells complained of. The remarkable thing was that every teacher and the principal pronounced the ventilation perfect. They stated that the conduct of the children as to lessons and behavior was noticeably better. No drowsy afternoons followed. Teachers stated they were as fresh at the close of the day as in the morning. Colds and coughs nearly disappeared. No contagious disease developed during the six weeks' trial, altho influenza was epidemic at this time. On several occasions a check was made on attendance, and not an absentee was reported on account of sickness.
>
> —E. S. Hallett, 1920

> A teacher should be a good housekeeper.
>
> —Kansas Department of Education, 1916

> The transformation of the United States in the late nineteenth and early twentieth century into a schooled society may be considered both as a part of the expansion of urban services and as an urban service that actually facilitated technological change.
>
> —Historian Eric Monkkonen, 1988

LIKE THE SALESPEOPLE OF DOHERTY'S FIRM, educators and home builders spoke of gas and electricity in terms of cleanliness, comfort, convenience, and economy of operation. However, each of these groups tailored its message to its own particular audience of clients, customers, or students. Because educators dealt each year with thousands of students who would eventually make countless decisions about gas and electric installations, it makes the most sense to explain their efforts first.

Early in 1916, editors of the *American School Board Journal* reported on construction and remodeling programs completed by local districts during the previous year. Regardless of region or size of the district, the emphasis of school officials and of the engineers and architects they had employed was on the benefits to faculty and students of installing modern systems for lighting and heating buildings. At a small school located in Sioux Falls, South Dakota, students were breathing air "warmed to about 70 degrees in the coldest weather." One of the new buildings in Detroit featured "electric lighting fixtures . . . [that] give uniform and adequate service at all times." The Rosemont School in Radnor Township, Pennsylvania, included "a vacuum cleaning outfit," as well as being "fitted with all devices to make it a model from the standpoint of health and hygiene."[1]

Before 1900, educators had regularly instructed students in health and hygiene, but that instruction took place without reference to gas and electric appliances, which up to 1900 both school officials and most North Americans had judged emblematic of novelty and luxury. After 1900, however, school officials assumed responsibility for providing every student with an environment that was comfortably warmed, appropriately ventilated, and suitably lit. By 1915, moreover, teachers regularly offered instruction to men in the techniques of appliance repair and to women in up-to-date methods of gas cooking and electric sewing. During the first two decades of the century, then, educators turned to gas and electric equipment to help teach older subjects, such as health and hygiene, and to foster newer forms of learning in such areas as cooking and the maintenance of power plants. As part of teaching students about power plants or cooking, educators also reinforced the boundaries of gender (and class) that were beginning to surround appliance purchase and use.[2] Altogether,

1. A. A. McDonald, "One-Story Schoolhouses in Sioux Falls," *American School Board Journal* 52 (February 1916), 30; "Recent School Buildings in Detroit," *American School Board Journal* 52 (February 1916), 15; "At the Rosemont School in Radnor Township, Pa.," *American School Board Journal* 52 (January 1916), 23.

2. Historians have prepared a substantial body of literature focused on the efforts of educated Americans to improve the health and hygiene of others. For the period before 1900, see John C. Burnham, *How Superstition Won and Science Lost: Popularizing Science and Health in the United States* (New Brunswick, N.J.: Rutgers University Press, 1987), pp. 56–62; Nancy Tomes, "The Private Side of Public Health . . .," *Bulletin of*

educators taught skills and fixed standards for ordinary Americans regarding the consumption of light and heat that extended into the next decade and beyond.

School officials in Kansas City and Denver were members of that larger group of educators who were installing electric lights and other new equipment. Similar to their colleagues in many cities, they identified the introduction of gas and electric service with an opportunity to offer training that would enhance employment skills, increase consumption by householders, and further encourage moral and physical prophylaxis. The technological and vocational high schools in Kansas City and Denver had been created with several of these purposes in mind. Consequently, vocational and technological educators were among the first to provide a fully developed articulation of the pedagogic advantages of abundant light and warm and rapidly circulating air.[3]

---

the History of Medicine 64 (Winter 1990), 509–539; Marilyn Thornton Williams, *Washing "The Great Unwashed": Public Baths in Urban America, 1840–1920* (Columbus: Ohio State University Press, 1991), pp. 22–40; Gwendolyn Wright, *Moralism and the Model Home: Domestic Architecture and Culture Conflict in Chicago, 1873–1913* (Chicago: University of Chicago Press, 1980), pp. 92–93; and Mary Corbin Sies, "The City Transformed: Nature, Technology, and the Suburban Ideal, 1877–1917," *Journal of Urban History* 14 (November 1987), 81–111.

3. The origins of technical and vocational education has emerged as a cottage industry among historians. Many find those origins among business executives. David B. Tyack, *The One Best System: A History of American Urban Education* (Cambridge: Harvard University Press, 1974), p. 189, contends that business leaders, such as J. P. Morgan, promoted development of technological and vocational courses of study. Similar in their findings are Julia Wrigley, *Class Politics and Public Schools: Chicago, 1900–1950* (New Brunswick, N.J.: Rutgers University Press, 1982); and David J. Hogan, *Class and Reform: School and Society in Chicago, 1880–1930* (Philadelphia: University of Pennsylvania Press, 1985). Equally, Maureen A. Flanagan, "Gender and Urban Political Reform: The City Club and the Woman's City Club of Chicago in the Progressive Era," *American Historical Review* 95 (October 1990), 1039–1040, determines that male industrial leaders favored vocational education in order to prepare reliable and productive workers, but that upper-income women, often members of the same families, favored vocational education to encourage youngsters to stay in school and eventually secure higher-paying jobs. Other historians find the origins of vocational and technical education among educators. Ira Katznelson and Margaret Weir, *Schooling for All: Class, Race, and the Decline of the Democratic Ideal* (New York: Basic Books, 1985), pp. 60, 68, conclude that educators rather than business or labor leaders took the initiative in promoting vocationalism. Walter Licht finds a still more complex set of relationships between changes in Philadelphia's economy and the vocational curriculum presented to the city's students. On the one hand, he determines that curricular innovations were not undertaken with a view toward serving industrial needs. In fact, he argues, "a new kind of work force was not needed." On the other hand, Licht determines that although business leaders favored vocational education, more important were administrators of

In 1900, the key pedagogical idea governing curricular developments at Denver's Manual Training High School was "to bring into close relationship *knowing and doing.*" In practice, the technological high schools provided a broad offering of courses in such areas as Latin, English, and advanced mathematics, along with drawing and shop courses in which the emphasis was on doing. This broad-based curriculum distinguished technological schools from vocational schools. In terms of gas and electricity, the orientation at Manual was in part theoretical and in part on the techniques of gas and electric utilization. Portions of the theory and the practice taught also depended on the student's gender. In brief, the technological high school taught the intricacies of gas and electric operations to male students and alerted women to their potential in the home.[4]

By the turn of the century, the science program at Manual prescribed the study of the dynamo and motor for young men. During their junior year, they took field trips to powerhouses to learn firsthand. For women, their training focused on domestic science, which included economy in food preparation and the physiology of the digestive process, along with practice in the details of specialized cooking. Women were also instructed in "fruit cookery . . . fish, meats, soups, and simple desserts." By spring, the emphasis shifted to "care of the dining room, table laying and serving, sanitation, ventilation, invalid cookery [and the] chemistry of cleaning." But more to the point in this context, students acquired their cooking skills with a gas range as well as Aladdin oven, twelve small gas heaters, and a coal range.[5]

By 1905, Kansas City's Manual Training High School offered a similar program of academic study and application. During their sophomore year, students selected from among courses in Latin, Ger-

---

the public and Catholic schools who were offering a "tangible education" as part of efforts to boost enrollments and bureaucratize and professionalize school systems. See Walter Licht, *Getting Work: Philadelphia, 1840–1950* (Cambridge: Harvard University Press, 1992), pp. 57–87, including quotations on p. 87. I want to thank Nina Lerman for calling this book to my attention. Finally, see Lerman's "'Of Practical Value': Technical Education in the Philadelphia Public Schools, 1880–1910," a revised version of a paper presented at the Annual Meeting of the Society for the History of Technology, Cleveland, Ohio, October 1990. Lerman contends that historians of education, and especially of technological and vocational education, have overlooked course content in favor of interest-group politics.

4. Manual Training High School, Denver, School District Number One, Arapahoe County, Colorado, *Courses of Study, Requirements for Admission, General and Specific Information* (1900), p. 10.

5. Ibid., pp. 14–15, 25.

man, botany, and American literature, as well as mechanical drawing, dressmaking, and domestic economy. School officials also created "a thorough course in drawing and the application of tools for boys and a course in domestic art and science for girls." Not even a course in elocution could escape the pressure to be practical and to conform to ordinary expectations about men and women. "Many a young man," argued an official in the school's announcement for 1905, "owes his puny salary to his inability to express what he knows gracefully and effectively."[6]

Naturally, instruction in the conceptualization and application of gas and electric systems was part of those multiple emphases on theoretical insights, practical study, and agreed-upon roles for men and women. As one illustration, the senior-year course in steam and electricity, for boys, assumed an understanding of "the theory of steam and the laws of combustion," along with "the laws governing electrical phenomena." In turn, students acquired direct information about steam and electricity by studying the actual operations of the school's heating and lighting plant. During the fall, students and instructors studied equipment in the boiler room. Thereafter, they moved to the engine room where the current was generated, controlled, and measured. By May, attention turned to an examination of the electrical, heating, and ventilating systems. In addition, students toured the electric plant located in downtown Kansas City, heard lectures by "practical engineers," conducted experiments, and learned to "make the calculations and drawings for dynamos and motors."[7]

The course in domestic science, for women, also called on students to make connections between the broader principles of science and the more prosaic matter of acquiring skills in meal preparation and house cleaning. Perhaps as a counterpart to the work being undertaken by many of the boys in the theories and techniques of steam and electricity, young women studied "the digestive organs, digestion, absorption, and circulation." With those materials in hand, students learned about "diet from infancy to old age . . . ," which led in turn

6. Manual Training High School, Kansas City, Missouri, *Course of Study for Admission: General and Special Information* (Kansas City, Mo., 1905), pp. 7, 45. Before 1900, educators had taught craft skills to young men and domestic skills to young women. The advent of gas and electric appliances created new opportunities for teaching. For the development of technical education according to class, race, and gender in Philadelphia during the nineteenth century, see Nina Lerman, "From 'Useful Knowledge' to 'Habits of Industry': Gender, Race, and Class in Nineteenth-Century Technical Education" (Ph.D. dissertation, University of Pennsylvania, 1993). Again, I want to thank Dr. Lerman for sharing her work with me.

7. Manual Training High School, *Course of Study*, pp. 39–41.

to a "study of menus and serving." The "cooking laboratory," as the room equipped with ovens and stoves was called, was located in the school's "shop wing." In conjunction with nearby rooms devoted to pattern-making and sewing, the cooking laboratory served as a crossroads at which educators located theory, technique, and the division of labor by sex.[8]

The curriculum at vocational high schools—as distinct from the technological schools—never included study of such subjects as ancient and foreign languages or the findings of the physical and biological sciences. By 1914, instruction in Denver's vocational high school emphasized the acquisition of skills judged useful in servicing electric and gas equipment and relevant to the distinct needs of young men and women. For instance, the electrical construction course included instruction in "building, winding, testing and repairing of dynamos, motors, and other electrical apparatus," as well as work on the installation of electrical wires for lights, bells, and so forth. "Any boy who successfully completes this course," ran a report in a city publication, "will have . . . such training as will enable him to become a skilled workman and even an overseer."[9]

Instruction for women in vocational high schools followed a second track. In Kansas City and Denver, courses such as mathematics and drawing were taken together with men, but thereafter women studied the applied aspects of domesticity. Home economics, announced a local publication, was "one of the strongest subjects which the girls are offered." Young women studied food preparation and service along with house cleaning and care, including care of the plumbing and ventilating equipment. Work in the "domestic science laboratory" with its many gas stoves was part of the routine of instruction in the home economics curriculum. School officials recognized that the "life work" of their female students might take place "in the home or office." Similar to the male-oriented courses of study, the emphasis overall at the school was on preparing students for employment as skilled workers who would eventually win promotion to the lower

8. Ibid., pp. 87–91. The contents of home economics courses in Denver and Kansas City cohered with those in the field nationally. See Isabel Bevier, *Home Economics in Education* (Philadelphia: J. B. Lippincott Company, 1924), p. 189, for the observation that "the work in food has long been taught from two standpoints: (1) that of the cooking school, and (2) that of the scientist." See also Margaret W. Rossiter, *Women Scientists in America: Struggles and Strategies to 1940* (Baltimore: Johns Hopkins University Press, 1982), pp. 67–70, who reports that home economics at the collegiate level was practical in its orientation, scientific in its method, and oriented exclusively toward women. I want to thank Ruth Schwartz Cowan for bringing these volumes to my attention.

9. "Denver's Technical High School," *The City of Denver* 2 (April 11, 1914), 11–12.

rungs of management. "The average girl," the reasoning went, "ought, if she chooses, soon to be able to supervise the work of others in ordinary industries and arts and to command good wages."[10]

Although pamphlets and bulletins published by school officials provide an overview of course contents, less certain is the precise number of students who actually studied digestion or lighting and who participated in the acquisition of practical skills in front of dynamos and stoves. Nonetheless, two sets of figures provide a crude measure of the possible number of persons exposed to the theory, practice, and gender of gas and electric machinery. First, between 1898 and 1905, more than 800 men and women graduated from Kansas City's Manual Training High School. They were among the best prepared persons in the city for dealing with the production and consumption of gas and electricity. As a second measure of the potential audience for this type of instruction, by 1912 the domestic science classroom in one of Denver's vocational schools had space for as many as thirty-two women at a time. In 1915, an official of the Denver Gas & Electric Light Company was able to report that "every [female] Denver high school student has an opportunity to take a course of domestic science with the full gas equipment."[11] By 1915, educators in both cities had systematized methods for diffusing knowledge about the application of gas and electricity to large numbers of students.

Focus on one of those students allows a fuller grasp of the connections between a course of study in a technological high school and the diffusion of knowledge about gas and electric appliances. In 1897, fourteen-year-old Roy G. Munroe enrolled in Denver's Manual Training High School. Oil lamps lit his parents' home. Munroe's mother cooked on a coal stove. At Manual, Munroe learned the rudiments of gas and electric operations. In 1901, following graduation, he took a job as a meter reader with the new Denver Gas & Electric Company, as did several of his classmates. Within a few years, company officials promoted Munroe to sales, where he brought knowledge of the use of gas and electricity to housewives and business leaders.[12]

10. Ibid.

11. Manual Training High School, *Course of Study*, pp. 107–113; Rufus G. Gentry to *Gas Age* 35 (May 1, 1915), 462.

12. Munroe Scrapbook. Munroe later emerged as an executive of the Denver Gas & Electric Company, suggesting his lack of typicality. See also Licht, *Getting Work*, pp. 87–97, for the argument that vocational education failed to prepare students for the details of industrial work. Instead, Licht contends, employers preferred to train students in those details. Nonetheless, reports Licht, the fact that students cared so little about formal education raises doubts about the ability of historians to link school

Between 1900 and 1915, educators exposed increasing numbers of students to the rigors and opportunities associated with the study of gas and electricity for home and office. The most plausible assumption is that most of these students were not employed in the gas and electric fields; they were simply better informed as consumers, whether homemakers, employees, or supervisors. Still other students, such as Roy Munroe, were employed precisely because of their knowledge of gas and electric applications. Those students—as electricians, plumbers, home economists, or sales personnel—devoted their careers to servicing these new systems or to transmitting information about gas and electric applications to city residents.

Education in the public schools about gas and electricity extended beyond the realms of practical applications and asserted connections to physics and physiology. Beginning around 1900, educators also began to create linkages for themselves and their students between such equally practical and theoretical matters as classroom heating and lighting, on the one hand, and learning, classroom deportment, and disease transmission, on the other. Health and hygiene had long been taught, but now educators began to attach those topics to gas and electric equipment.

Magnificent science was once again at the core of these initiatives. This science was not simply the idiomatic "science" of the late nineteenth century, which was often a word for anything not well understood. "Science" in the early years of the twentieth century suggested a method by which educated Americans might shape their environment and the environments of their neighbors. Between 1900 and 1920, argues historian John C. Burnham, professional men and women including members of the clergy, social workers, and educators embraced this new definition. "Science had justified both extravagant hopes and men's beliefs in their own power to change the world." In turn, science, as they conceived it, included the applicable products of technology and medicine. After 1900, then, as directors of the nation's school systems embraced the findings and methods of science, they also endorsed the germ theory of disease and the concomitant obligation to offer instruction in public health. In order to fulfill that obligation, educators in many cities began to fix the rules for classroom heating, air circulation, and lighting with a view toward positively affecting the health and sight of students. At a minimum,

attendance and course contents with a subsequent improvement in their problem-solving skills or their sense of work discipline.

improved facilities would offer a standing lesson to each on the relationships among air, heat, light, and hygiene. With luck, as Burnham puts it, "students [would] learn about and master their immediate environments." Increasingly, educators judged gas and electric machinery (as well as modern plumbing) requisite to controlling that environment.[13]

Between 1900 and 1915, the reform idea linking science, health, mechanical heating, and electric lighting had reached the school districts of the mid-continent and Rocky Mountain regions. "Confinement in an improperly heated and poorly ventilated room," contended Kansas educators in 1916, "saps the vitality and undermines the health of both teacher and pupils and decreases their working power." Air had to circulate freely, they contended, at a temperature between 68 and 70 degrees. "There should also be provision for bringing in the fresh air from outside the building, heating it, and removing the foul air." This "foul air," as they labeled heated air that students had breathed, had to be "drawn off through the foul-air vent."[14]

Circulation of air that was heated to a fixed and comfortable temperature was only the first step in the process of improving the health of each student. Modern lighting systems held out the promise of improved sight. "Defective vision among school children," which Kansas educators blamed on "improperly lighted schoolrooms," was remediable through redesign and relocation of the classroom. "Light from only one side of the schoolroom is best," they contended, "and that . . . light should come from the left of the pupils." The total size of windows in each classroom, moreover, "should be not less than one-fifth the floor area." At the same time, "sanitary reasons" led the authors of this set of requirements and suggestions to insist on the admission of "direct sunlight . . . into the room" during a portion of the day. By fixing the heating and lighting requirements for each classroom, Kansas educators were also setting a clearly defined physical framework judged crucial to general and health education.[15]

Recommendations of state officials could only fix broad standards

13. John C. Burnham, "Essay," in John D. Buenker et al., eds., *Progressivism* (Cambridge: Schenkman Publishing Company, 1977), pp. 18–19; Burnham's *How Superstition Won and Science Lost,* pp. 183–184. See also "Hygienic and Physical Education in the Kansas City Schools," *School and Society* 2 (October 2, 1915), 489–490.

14. State of Kansas, Department of Education, *Standard Rural Schools: Requirements and Suggestions of the State Board of Education* (1916), pp. 8–9.

15. Ibid., pp. 6–7. See also Burnham, *How Superstition Won and Science Lost,* pp. 56–62, for the methods of health popularizers, including educators, in transforming generalized goals into everyday practice.

for classroom heating and lighting. Beginning after 1910, educators in Kansas City and Denver turned increasingly to consultants and other experts who could bring the appearance of mathematical certainty to recommendations concerning new heating and lighting systems and expected improvements in student learning and behavior. In turn, these experts found in school work yet another opportunity to express their own training and experience in the politics of numbers.[16]

Part of the work of measurement in the schools fell under the control of directors of newly created bureaus of research and efficiency. Between 1911 and 1914, school administrators in nine cities added those units to their central administrations. In 1914, Kansas City's bureau began operations under the supervision of George Melcher, a former assistant superintendent of the Missouri school system. By 1915, Melcher had studied the costs of heating seventy-seven elementary schools and five high schools. Although the stated emphasis in this study was on reducing the costs of coal, creation of a comfortable temperature in every building remained the overriding goal.[17]

Rather than create a research bureau staffed by their own experts, educators in Denver turned to Lewis M. Terman, an associate professor of education at Stanford University and later famous in the field of mental testing. Terman's report, delivered in 1916, assembled a number of contemporary ideas about light, heat, hygiene, and good education. Most of Denver's school buildings, Terman complained, were old. Worse yet, architects not familiar with contemporary standards of heating and lighting had designed them. Although the exterior appearance of those buildings "have not displeasing proportions," Terman reported that "few of the school rooms are well lighted [and] many of the heating systems are primitive." Terman also found that "the ventilation is unsatisfactory in at least half of the buildings," with the result that "dusty and parched air is carried into the classrooms." Just as much in need of improvement were the lavatories, which Terman described as "usually dark, often cold and sometimes

16. For the politics of expertise in the field of urban transportation, see Paul Barrett and Mark H. Rose, "Street Smarts: The Politics of Empiricism in Shaping Urban Technology, 1900–1950," in Kathleen Neils Conzen, Michael H. Ebner, Russell Lewis, and Eric H. Monkkonen, eds., *American City History: Modes of Inquiry* (Chicago: University of Chicago Press, in press).

17. "The Kansas City Bureau of Research and Efficiency," *School and Society* 2 (July 24, 1915), 127; Kansas City, Missouri, Public Schools, Bureau of Research and Efficiency, *Report of the Bureau* (1914–1915), 1 (February 1916), 80–89.

foul-smelling." So unsatisfactory were several of the older buildings that Terman believed they "should be replaced with the least possible delay."[18]

Terman argued from numbers when making his proposals to Denver administrators for improving the lighting and heating of their schools. He had measured lighting in classrooms and lavatories in terms of ratios of natural to artificial illumination. In order to achieve a proper level of illumination for each student, Terman asserted that educators had to take account, as they did in Kansas, of the ratio of floor to window area. Improvements to the heating systems of older schools also reflected the application of an agreed-upon set of numbers. Educators had already settled on a temperature of 68 to 70 degrees as desirable. Temperatures above that level, argued Terman, led to "headaches, a feeling of malaise, and nervousness."[19]

About all that remained for study and determination was the appropriate volume of air circulation and the location of air ducts. No longer was it adequate for the ventilating engineer to supply "thirty cubic feet of fresh air per minute per child." Contemporary standards, argued Terman, required that "the body is bathed in perceptible air currents." The precise location of inlets and outlets to supply those currents, Terman observed, should be determined "with reference to the shape of the room and the location of outside walls." But designers could begin with the knowledge that "the proper distance from the floor [of the fresh air inlet] is about eight feet." Inlets set nearer the floor "become receptacles for dust and dirt, which is blown into the schoolroom and breathed into the lungs of the children."[20] Terman's study and report affirmed the idea that it was possible to achieve improvements in the health of children through exact mea-

18. Lewis M. Terman, *Report of the School Survey of School District Number One in the City and County of Denver, Part V: The Building System and Medical Inspection* (Denver, Colo.: School Survey Committee, 1916), pp. 6–7. See also Ernest Bryant Hoag and Lewis M. Terman, *Health Work in the Schools* (Boston: Houghton Mifflin Company, 1914). Henry L. Minton attributes Terman's interest in school hygiene to an attack of tuberculosis in 1900; see his *Lewis M. Terman: Pioneer in Psychological Testing* (New York: New York University Press, 1988), p. 16.

19. Terman, *Report of the School Survey,* pp. 28–34.

20. Ibid., pp. 28–34. For an account of the efforts of contemporary, upper-income women to purify the environment and alter behavior, see Suellen M. Hoy, "'Municipal Housekeeping': The Role of Women in Improving Urban Sanitation Practices, 1880–1917," in Martin V. Melosi, ed., *Pollution and Reform in American Cities, 1870–1930* (Austin: University of Texas Press, 1980), pp. 173–194; and for an account of the relationship between those seeking to clean cities and others seeking to improve health, see James C. Whorton, *Crusaders for Fitness: The History of American Health Reformers* (Princeton: Princeton University Press, 1982), pp. 144–145.

surement of light and heat and a consequent restructuring of the built environment.

Terman's report was one of many prepared by educators and physicians who were seeking to upgrade the physical environments of children. Often, those physicians were also educators, having served as members of school boards. Experience and study guided their observations. As participants in earlier and contemporary reform efforts, educators and physicians had witnessed substantial improvements in the well-being of children and adults, including marked reductions in deaths caused by diphtheria, scarlet fever, measles, and tuberculosis. After 1900, educators throughout the nation were learning that modern systems of heating and lighting could advance older goals of improved health, learning, and deportment. For example, introduction of abundant lighting that was properly aligned with walls, windows, and desks held out the realistic promise of improved vision and improved reading, leaving fewer youngsters to grow to adulthood with poor sight and illiterate. "The inconvenience and misery" of eye defects, reported two researchers in 1913, "are indeed out of all proportion to the ease with which they can be removed or prevented."[21]

After World War I, these studies and the health crusade that promoted them began to inform heating and lighting arrangements in newly constructed schools. Review of the specifications prepared by school architects and engineers allows an examination of the diffusion of ideas about health and hygiene and their perceived links to precise standards of light and heat. The design standards and blueprints of school architects and engineers represented the health crusade stripped to the bone.

21. Stuart H. Rowe, *The Lighting of School-Rooms: A Manual for School Boards, Architects, Superintendents, and Teachers* (New York: Longmans, Green & Company, 1904); Ernest Bryant Hoag, *The Health Index of Children* (San Francisco: Whitaker & Ray-Wiggin Company, 1910); May Ayres, Jesse F. Williams, and Thomas D. Wood, *Healthful Schools: How to Build, Equip, and Maintain Them* (Boston: Houghton Mifflin Company, 1918); E. H. Lewinski-Corwin, "The Public Health Movement," in Louis W. Rapeer, ed., *Educational Hygiene from the Pre-School Period to the University* (New York: Charles Scribner's Sons, 1915), pp. 37–41; Francis Willston Burks and Jesse D. Burks, *Health and the School: A Round Table* (New York: D. Appleton and Company, 1913), p. 20. Modern scholars affirm many of the insights gained by this generation of educators and physicians. Education about the transmission of disease, argue Samuel H. Preston and Michael R. Haines, was more important than income or diet in reducing the death rate of children. By 1925, they report, the children of teachers and physicians "had relative levels of . . . mortality that were 64 percent and 66 percent of the national average respectively compared to their values of 100 percent and 94 percent in 1895." See Preston and Haines, *Fatal Years: Child Mortality in Late Nineteenth-Century America* (Princeton: Princeton University Press, 1991), p. 209.

Quantification, expertise, and health converged in designs for the new C. S. Morey Junior High School in Denver. In March 1920, William E. and Arthur A. Fisher presented specifications for heating, cleaning, and ventilating the new school. The Fishers, who were brothers, architects, and partners, had years of experience in designing schools, other public buildings, and the homes of Denver's wealthiest residents. Their specifications for the Morey school appeared mostly in numeric form. For instance, contractors were to install several large fans in order to maintain air circulation. Each fan had to "have a rating of 43,200 cubic feet per minute, with outlet velocity of not more than 1,800 feet per minute with fan running at not more than 151 R.P.M." Similarly, installation of "the Johnson Service Company system of temperature control" in every room was intended to create a temperature conducive to improved studies and behavior, even allowing for an outside temperature as low as 10 degrees below zero.[22]

Not every area of school design was subject to numbers. Because no standard existed by which to declare an odor obnoxious, architects such as the Fishers were more tentative in making assertions about the installation of costly environmental-control equipment in locker rooms and lavatories at the Morey school. Lockers for the new school had not yet been designed, and the architects recommended perforating the top and bottom of each locker door as part of the ventilating plan. The idea was that "any odors accumulating from damp clothes will be drawn up through the lockers and there will be no odors in the room." Otherwise, argued the Fishers, the ventilating plan, including the circulation of warmed air, "will be useless."[23]

On-site inspections by the Fishers of the boys' lavatories in schools located in the East and the Midwest also awakened concerns about additional sources of odor, in this case the odor created in urinals. From the report of a janitor in a St. Louis school, the Fishers determined that the toilets in that building "were absolutely free from odors." Design of the urinals had helped create an odor-free room. The key feature in controlling odors, however, was "that a direct and positive ventilation was obtained at all times." One question the Fishers put before members of the Denver school board was whether to

22. "Heating, Ventilating, and Vacuum Cleaning Specifications for the Junior High School, 14th Avenue and Clarkson Street, Denver, Colo.," March 1920, pp. 9–10, 13–14, 17; and W. E. and A. A. Fisher to F. H. Cowell, September 8, 1920—both in Fisher Architectural Records Collection, 1897–1978, Denver Public Library (cited hereafter as Fisher Records).

23. W. E. and A. A. Fisher to J. J. Hall, November 23, 1920, Fisher Records.

spend an additional $433 to purchase the type of urinal installed in St. Louis. The second question was whether the board wanted "direct ventilation for these urinals" as opposed to the ventilation "we have already provided."[24]

Beginning around 1920, architects included areas that were low in traffic and low in status, such as lavatories and closets, in plans for environmental improvement. The idea, as the Fishers reported to members of the school board, was "to have these toilet rooms absolutely satisfactory." During the 1920s, the efforts of educators and architects such as the Fishers to improve "the comfort and health of the school children" were resulting both in a redesigned curriculum and in the redesign of every room in the school.[25]

By about 1920, leaders of the health and hygiene crusade in the nation's schools had enjoyed considerable success in brightening classrooms and providing a more uniform level of heat and ventilation. In a few schools, architects had even begun to address the newly discovered problem of odors. After 1920, however, successes of these sorts began to appear insufficient. Rooms ought to be brighter; ventilation systems needed to run faster; odors, a sure sign of bacteria, were to be eliminated. During the next few years, these concerns found more formal expression in two arenas. First, advocates of health and hygiene in the classroom cooperated with architects and engineers on the national level in writing codes to upgrade technical standards, especially for school lighting. In turn—and this was the second arena—many of those same advocates for programs of improved health and hygiene in cities such as Denver promoted massive school construction programs. Upgraded codes and rapid construction of up-to-date schools went together. The idea was that sooner rather than later the study habits and health of every child would benefit from exposure to an enhanced school environment.[26]

24. Ibid.; W. E. and A. A. Fisher to the School Board, February 15, 1921, Fisher Records.

25. Ibid. Increasingly, notions of good taste, personal comfort, and public decency—at least as civic leaders, engineers, and architects interpreted those concepts—included high-speed fans to eliminate odors and reduce the chances of transmitting germs. For the introduction at considerable expense of fans into Kansas City's public restrooms, see "Public Comfort Stations at Kansas City, Mo.," *Domestic Engineering* 62 (January 4, 1914), 1–7; and Ralph R. Benedict, "Municipal Bath-House in Business District," *American City* 18 (January 1918), 46–47. For earlier efforts of engineers and other experts in a number of cities to purify air, see R. Dale Grinder, "The Battle for Clean Air: The Smoke Problem in Post–Civil War America," in Melosi, *Pollution and Reform in American Cities,* pp. 83–101.

26. For the concepts of expertise, national engineering standards, and local administration, see Bruce E. Seely, *Building the American Highway System: Engineers as Policy Makers* (Philadelphia: Temple University Press, 1987), pp. 71–135. For the politics of

In 1918, school engineers, manufacturers, and school administrators had prepared a code for schoolroom lighting. By 1921, the states of New York and Wisconsin as well as several cities had adopted the code. According to a report in an engineering journal, adoption of the code had "led to marked improvement in school lighting throughout the country." During the early 1920s, however, a coalition of experts including school architects, educators, and electrical engineers judged the code inadequate. In 1924, they prepared a revised code. Authors included items such as "definite requirements under the glare rule [and] a limiting ratio of maximum intensity to minimum intensity." According to a report in an electrical engineering journal, "the new code is more specific." Because engineers, architects, and educators often accepted the inherent value and neutrality of numbers built into technical codes, it became possible for members of those groups to coalesce around a common body of data and then to use that data to launch efforts in local political arenas to secure approval for the construction and remodeling of school buildings.[27] Codes facilitated the politics of diffusion.

In the design and construction of schools, however, implementation of even the most precise technical standards rested on local initiatives. Beginning in 1919, business and political leaders in Denver allied with school administrators in an effort to persuade voters to approve a bond issue to fund a program of school-building. "Businessmen of Denver would not tolerate such conditions as confront our school children," claimed an advertisement dated September 18, 1919, placed in the Chamber of Commerce weekly by members of the School Bond Election Committee. "School children go each day to crowded schools with badly ventilated . . . rooms," they argued, and "too many are ill every year from polluted air—air you'd not tolerate in your offices." In October 1922, Denver voters approved a bond issue for $6.1 million to finance construction of twenty-seven schools. To ensure proper lighting and heating, as determined by professional

industrial standards, see Bruce Sinclair, "At the Turn of a Screw: William Sellers, the Franklin Institute, and a Standard American Screw Thread," *Technology and Culture* 11 (January 1969), 20–34; and David F. Noble, *America by Design: Science, Technology, and the Rise of Corporate Capitalism* (New York: Oxford University Press, 1979), pp. 69–83, where the emphasis is on corporate hegemony in the determination of technical standards.

27. "'Code of Lighting School Buildings' Now a Standard," *Electrical World* 84 (September 27, 1924), 702. See also U.S. Bureau of Labor Statistics, "Code of Lighting: Factories, Mills, and Other Work Places," *Bulletin* 331 (April 1923), 26–28, for the contention that enhanced lighting of factories would lead to fewer accidents, improved production, neatness, and lower insurance rates.

standards, members of the school board hired a heating and lighting engineer and two ventilating engineers. First, they studied local coal supplies and heating equipment. Next, they recommended installation of electric heating units in windows. According to a report published several years later by the school board, these engineers liked "the unit heater and ventilator, [which] takes the air directly from the outside, . . . heats it while it passes over and through steam coils and forces it by means of electric fans into the room." In 1928, following installation of these units, the author of a report prepared for the board was able to determine that "vitiated air leaves the classroom [and travels] into the attic and out of the building."[28]

By 1921, combined enrollment in the public schools of Kansas City and Denver totaled nearly 110,000 students.[29] Teachers in Kansas City and Denver regularly informed students about the advantages of measured quantities of light and heat for improved study and health. A smaller number of students also acquired specialized knowledge of gas and electric principles and applications in domestic science and shop courses, leading for some to technical and sales positions and for many to an awareness regarding the advantages of proper lighting and appropriate heating and ventilation in the home, office, and plant. Instructors also offered those general and technical lessons in buildings designed by architects and engineers accustomed to calibrating light, heat, and ventilation in precise form. Mathematical formulas guided illumination levels; uniform heating at 68 to 70 degrees served as a standard; and school architects and engineers designed new facilities to suppress odors and deliver clean and humidified air in a constant flow. In each of these areas, the appropriate roles of men as supervisors of machinery and of women as supervisors of appliances were regularly reinforced.

For all of the dramatic changes taking place in the pedagogy and physical environment of schools, historians nevertheless recognize that what students were taught may well have diverged from what

28. *Denver Commercial* 10 (September 18, 1919), 3 (advertisement); Denver Public Schools, *The Denver School Building Program* (1928), pp. 13–14, 19–20, 27–28, 30. See also G. L. Lockhart, *Public Schools: Their Construction, Heating, Ventilation, Sanitation, Lighting, and Equipment* (St. Paul, 1918), pp. 7–10, 12–13, 74–87; and E. S. Hallett, "A New Method of Air Conditioning in School Buildings: The Use of Ozone to Remove Odors and to Revitalize the Air," *American City* 22 (April 1920), 423.

29. Denver Public Schools, *Forty-Fourth Annual Statistical Report, 1946–1947,* p. 37; Kansas City, Mo., School District, *Report of the Superintendent of Schools of the School District of Kansas City, Missouri, for the Four Years from July 1, 1917 to June 30, 1921,* p. 139.

they perceived, remembered, understood, and practiced. Yet the study of gas and electricity within a built environment designed for warmth, ventilation, and bright lights did produce significant changes in the behavior and outlook of many students. Those changes took place along two lines.

First, it was undoubtedly the case that many youngsters learned the rudiments of electrical and mechanical operations in their city's high school. During April 1921, for example, students enrolled in Kansas City's technological and vocational schools participated in a local electrical trade show along with more than 100 other exhibitors. According to an account in a trade journal, the students' exhibits included kitchen appliances, electrically operated woodworking equipment, and "a complete sewing room in operation."[30] Perhaps many of those male students graduated into full-time positions as plumbers, electricians, and sales representatives for gas and electric firms. In those positions, they would have joined a growing number of technical specialists who were converting gas and electricity from the realm of esoteric science for the few into the realm of technology for a large number of urban residents.

Second, instruction in the public schools, and especially in the technological and vocational high schools, probably encouraged students to begin to identify hygiene, health, and environmental regulation with such mechanical devices as stoves, light bulbs, and electric window heaters. Rather than contend that schools taught a particular point of view or changed habits on the spot, it appears more reasonable that instruction in health and the presence of mechanical devices (such as fans to reduce odors) resonated with the outlooks of students already steeped in the culture of health and hygiene. During the 1920s and later, manufacturers produced a large number of devices for home and office aimed at reducing odors and germs. Leaders of those firms must have identified a growing market for machinery capable of "removing dust and air-borne organisms from the atmosphere."[31] As in other areas of life, health had a technological fix.

Educators, school architects, and school engineers also acquired new skills and outlooks as a result of their involvement with gas and

30. "Kansas City Electrical Show Successful," *Electrical World* 77 (April 30, 1921), 1011.

31. Arthur Cecil Stern, "Control of Air-Borne Bacteria," p. 73; H. C. Murphy, "Control of Air-Borne Dust and Smoke," p. 75; F. H. Munkelt, "Control of Odors," p. 76—all in *Architectural Record Combined with American Architect and Architecture* 86 (July 1939). For the connection in the minds of health and hygiene reformers between an improved environment and modern machinery, see Whorton, *Crusaders for Fitness*, pp. 164–165.

electric equipment. By 1900, as Burnham reminds us, every state legislature required schools to offer instruction in physiology and hygiene. Educators in Kansas City and Denver, along with their counterparts in virtually every city, regularly taught those subjects.[32] Up to 1900, however, those educators, like most Americans, enjoyed little firsthand experience with gas and electricity as applied to environmental regulation, vocational education, or preparation of students for the purchase of gas and electric appliances. After 1900, then, participation in the health and hygiene crusade encouraged educators to begin to revise curriculums and redesign buildings. Gas and electric equipment assumed a central place in these new buildings and curriculums, as educators stressed healthful cooking with gas, improved circulation of air and fewer diseases with forced-air systems, and improved reading and sight with electric lights located at proper angles to desks and windows. In the course of many years of shaping curriculums, designing buildings, and directing the installation of lighting and heating equipment, educators, engineers, and school architects enhanced and systematized their own knowledge. By 1924, practitioners were able to coalesce around a national code for school lighting, which amounted to a grant of political legitimacy for consensus technical practice. Educators were also capable of learning.

The introduction of gas and electric appliances also encouraged a more subtle alteration in the outlook of educators and their students. Not until after World War II would most of the students and educators in the nation's cities actually purchase the modern appliances on which instruction was offered. Nor would most install electric lights in their homes that met up-to-date standards of illumination or enjoy the uniform heating of a forced-air furnace. In reality, however, the relationship between building design, curriculum reform, and classroom education and the acquisition of electric and gas equipment was never a straightforward and short-term process of learning and then doing. Instead, students and faculty who participated in trade shows, cooked with gas on a school stove (and visited appliance dealers and department stores), and read about healthful living under bright lights were participating in a longer-term process described by Jean-Christophe Agnew as one of learning "a way of seeing." During the next few decades, whether as sales representatives for the gas and electric companies or as educators and students who were also consumers, they would make "countless contemplated purchases" in

32. Burnham, *How Superstition Won and Science Lost,* p. 55.

preparation "for every actual purchase."[33] Following World War II, those students and educators could begin to put into practice in their private lives the skills and outlooks acquired years before in the nation's public schools. As adults in the postwar United States, those skills, outlooks, and purchases also conformed to the expectations of gender created in schools and elsewhere a decade or two before.

Educators comprised only one portion of a much larger and heterogeneous group of urban Americans who carried practical information about the advantages of light and heat to students, neighbors, colleagues, customers, and clients. After 1900 or so, sales personnel for the great gas and electric firms and home builders catering to wealthier clients also assumed important roles as agents of diffusion. In particular, home builders such as J. C. Nichols in Kansas City were alert to a growing market among their upper-income clients for enhanced levels of health and hygiene that were to follow installation of modern lighting, heating, and ventilating equipment. Ordinary sales personnel, such as Roy G. Munroe, catered to Denver's wealthiest householders as well as to those who were much less fortunate. Munroe and Nichols directed and participated in large organizations that were increasingly part of a political economy that was organized around national flows of capital, ideas, and personnel. In their day-to-day work, however, each had to persuade residents of their city one at a time to purchase modern gas and electric equipment. Munroe and Nichols focused on the advantages of gas and electric appliances for members of an urban culture who were increasingly seeking cleanliness, comfort, and convenience, especially for women. The net result of their efforts up to 1940 was to produce ecologies of technological knowledge and practice.

33. Jean-Christophe Agnew, "The Consuming Vision of Henry James," in Richard Wightman Fox and T. J. Jackson Lears, eds., *The Culture of Consumption: Critical Essays in American History, 1880–1980* (New York: Pantheon Books, 1983), p. 73. I want to thank John C. Burnham for bringing this essay and several other items cited in this chapter to my attention, and I also want to acknowledge Professor Burnham's helpful comments on an early draft.

CHAPTER FIVE

Two Salespersons and Ecologies of Technological Knowledge, 1920–1940

> The natural gas business today must be conducted along the lines of the modern public utility business. The public must be sold and we are now in the third phase or period, the selling period.
>
> —T. J. Strickler, Vice President and General Manager, Kansas City Gas Company, 1927

> The true molders of the twentieth-century city were not Arthur Farwell, Jane Addams, or Daniel Burnham, but Al Capone, J. C. Nichols, and the Levitt Brothers.
>
> —Historian Jon C. Teaford, 1986

AFTER 1900, THE LARGE CORPORATION became a central feature of the North American economy. In turn, executives of these corporations created extensive apparatus for designing and marketing new products. Likewise, leaders of gas and electric corporations directed vast productive forces and possessed considerable marketing savvy, yet they did not find it easy to persuade householders in distant cities to purchase modern appliances. Rapid diffusion of gas and electric appliances still required the services of local people who could bring the company's message from distant places and adapt it to local circumstances. Maintenance of technological momentum and the success of high-level executives required the services of local agents who interpreted national themes for residents of heterogeneous and fast-changing cities.[1]

1. For development of the managers who directed several of these new corporations, see Olivier Zunz, *Making America Corporate, 1870–1920* (Chicago: University of Chicago Press, 1990). For the period after about 1920, modern historians have no

J. C. Nichols and Roy G. Munroe were two of those local persons who presided over large organizations dedicated to promoting gas and electric service. Nichols directed construction of homes for leading Kansas City families based on the idea of achieving environmental control through the installation of modern appliances in homes built among trees and parks. "Qualitatively," reports Kenneth T. Jackson, Nichols was "the most successful American developer." Munroe, a graduate of Denver's Manual Training High School and for many years a sales representative with Henry Doherty's Denver Gas & Electric Company, was by the early 1920s a sales manager as well as salesperson in the firm's new business department. During the course of a lengthy and successful career in sales and sales management, Munroe connected knowledge of gas and electric appliances to the tastes of Denver residents, linking stoves, irons, and furnaces to the promise of enhanced comfort and convenience.[2]

On the surface, Nichols and Munroe stand in stark contrast to one another. By the early 1920s, Nichols, a wealthy and prominent developer in Kansas City, had achieved a national reputation as a leader in city planning and in the design and construction of fashionable homes. Munroe, a salesperson and then a sales manager, was active in local organizations but was and remains one of those anonymous Americans who staffed a specialized bureaucracy for one of America's large corporations. Nichols, moreover, exercised considerable discretion in the design of his homes, helping to determine the standards

---

doubt that the adjustment of supply, including the introduction of new products and technologies, was a matter of executive decision-making and implementation through organizations staffed by experts. The standard work in this area is Alfred D. Chandler Jr., *The Visible Hand: The Managerial Revolution in American Business* (Cambridge: Belknap Press of Harvard University Press, 1977); Thomas P. Hughes, *Networks of Power: Electrification in Western Society, 1880–1930* (Baltimore: Johns Hopkins University Press, 1983), p. 140, describes the assembly of large-scale electrical operations along with supporting organizations—governments, universities, and so forth—in terms of an apt metaphor, that of momentum. "The systematic interaction of men, ideas, and institutions," he informs us, "led to the development of a supersystem—a sociotechnical one—with mass movement and direction." Although Hughes is describing an earlier period during which electrical engineers, politicians, inventors, and investors coalesced around high-voltage alternating current systems, the concept is valuable for later decades as educators, home builders, sales personnel, and many others began to attach their careers and often their cash to the advantages of gas and electric service.

2. Kenneth T. Jackson, *Crabgrass Frontier: The Suburbanization of the United States* (New York: Oxford University Press, 1985), p. 177. For a case study of a technologically oriented person as both entrepreneur and mediator among large firms, see W. Bernard Carlson, *Innovation as a Social Process: Elihu Thomson and the Rise of General Electric, 1870–1900* (New York: Cambridge University Press, 1991).

of good taste for his clients and for subsequent generations of Kansas City's residents. Munroe, by contrast, supervised programs that had been set in place by executives in New York and by his superiors in Denver. In all, our own understanding of wealth, influence, and prestige suggests that Munroe and Nichols may serve merely as vehicles for comprehending substantially different facets of business experience, urban change, and leverage in two sectors of the American political economy.

Yet for all of their obvious and apparent differences, Munroe and Nichols provided similar services to residents of their cities. They emphasized cleanliness, comfort, and convenience inside of a controlled environment; they appealed to householders anxious about economic and urban upheavals; and they pitched the sale of their homes and appliances with a view toward changes in the family and in gender relationships. Ultimately, these two salespersons helped shape the aesthetic sensibilities, the built environment, and even the technological awareness of those possessing incomes in about the top one-third of their cities. Both helped shape ecologies of technological knowledge. Roy Munroe and J. C. Nichols provided services that were not so much comparable as complementary.[3]

By the early 1920s, the Denver Gas & Electric Company had emerged as a large and successful corporation. In 1910, Henry Doherty had purchased the company from Emerson McMillan, initiating a nationwide program of financing and managing electric and gas corporations. In 1920, a Doherty publicist described the organization as "among the leading financial houses of the country." In 1924, as part of a strategy of combination that was increasingly common in the gas and electric field, Doherty merged his company in Denver with other gas and electric firms in the Rocky Mountain west under the name "Public Service Company of Colorado."

During the 1920s, Doherty connected a number of his gas companies to long-distance pipelines, providing natural gas to numerous householders and manufacturers in cities along the right-of-way. Electric operations grew rapidly as well. By 1935, Doherty's Cities Service Company owned 183 electric companies located in eighteen states and was one of the largest electrical operators in the United States.

3. See Jon C. Teaford, "Finis for Tweed and Steffens: Rewriting the History of Urban Rule," *Reviews in American History* 10 (December 1982), 147, for the contention that the history of urban government is a history of those who "struggle[d] to provide vital services for millions of Americans demanding unprecedented levels of comfort and convenience."

From corporate headquarters at 60 Wall Street in New York, Doherty's senior executives helped shape the budgets of their local companies and the pace of economic development in several regions of the United States. During an era in which business executives became self-conscious about the importance of public relations, however, a Doherty publication issued in 1928 celebrated "industrial and business giants . . . in the service of humanity."[4]

For all of the ability of senior officials in New York to anticipate and perhaps shape markets, they understood that certain decisions about sales activities in Denver still had to be made at the local level. In Roy G. Munroe, managers recognized a person whose energy, intelligence, and accomplishments would soon earn him promotion to higher levels of responsibility within the company. In 1901, following graduation from Denver's Manual Training High School, Munroe joined Denver Gas & Electric as a meter reader. In 1904, after several years in that post and a year-long assignment in the collection department as an assistant bookkeeper, Munroe was promoted to sales. He joined Doherty's growing staff of representatives who called on householders to sell electric and gas lighting and appliances. In 1906, he was promoted to the rank of service supervisor, which was essentially a sales position but one judged more important by company officials. Munroe also participated in the required training programs that met several evenings each week and provided instruction in the details of electric and gas operations and sales techniques. In 1914, he enrolled in a correspondence course in industrial gas engineering that included laboratory work at the University of Denver. One portion of the course required students to prepare a plan to increase gas consumption among area residents. Munroe's proposal was to examine

4. William P. Strobhar, comp. and ed., *A Manual of Communities Served by Corporations Operated by the Henry L. Doherty Organization* (n.p.: Doherty's Men's Fraternity, 1920), p. 5; Cities Service Gas Company, *Building an Empire* (n.p., c. 1928), in Lou E. Holland Papers, University of Missouri at Kansas City; U.S. Federal Power Commission, *National Power Survey: Principal Electric Utility Systems in the United States, 1935*, Power Series No. 2 (Washington, D.C.: Government Printing Office, 1936), pp. 12–15, 22–25. For the organization of multistate holding and operating companies in the electrical field during the 1920s, see Hughes, *Networks of Power*, pp. 324–334; for the growing importance of public relations in American business, see Louis Galambos and Joseph Pratt, *The Rise of the Corporate Commonwealth: U.S. Business and Public Policy in the Twentieth Century* (New York: Basic Books, 1988), pp. 92–99; and for public relations as a method senior managers used to reshape a firm's internal operations by invoking the service ideal, see Roland Marchand, "The Corporation Nobody Knew: Bruce Barton, Alfred Sloan, and the Founding of the General Motors 'Family,'" *Business History Review* 65 (Winter 1991), 825–834.

current levels of consumption, to link that data to a study of Denver's growing population, and then to launch "an aggressive canvass" for new business.[5]

Between 1904 and the early 1920s, Munroe merged his knowledge of gas and electric appliances and consumer tastes to develop effective sales techniques. As the company introduced new products requiring extensive promotional efforts, Munroe was frequently the leading salesperson. Around 1905, he sold the first electric iron. In 1910, he sold the first automatic gas hot-water heater; and in 1915, he took first place in a contest to sell the largest number of electric irons. During 1919 and 1920, Munroe assumed an active role in persuading foremen, superintendents, and mechanics at railroad shops in Denver to substitute gas for coal in heating locomotive wheels for refitting. As late as 1916, according to Munroe's report in a trade publication, the company was selling 211,000 cubic feet per month of gas to railroad shops. By 1920, that figure had jumped to more than 2.5 million.[6]

Beginning in 1915, Munroe assumed managerial responsibilities along with his sales work. His first assignment was as superintendent of industrial and commercial gas. Munroe and his sales representatives concentrated on persuading hotel, railroad, and manufacturing executives to convert from coal to gas. Once customers had connected to company lines and begun using gas, Munroe and his representatives returned, touting the benefits of increased consumption. In 1922, officials promoted Munroe to the post of assistant commercial manager, which increased his level of responsibility in preparing and supervising promotional efforts. Between 1924 and 1935, Munroe served as gas new business manager, a position created to boost the sale of gas for industrial, commercial, and residential purposes, especially cooking and space heating.[7]

Naturally, Munroe played no part in the decision to create his new position or to launch massive campaigns to boost the sale of gas. Years

5. Munroe Scrapbook; *Who's Who in the Doherty Organization* (n.p., c. 1920), p. 27; Roy G. Munroe, "N.C.G.A. Course III," c. 1914, both in Munroe Files. See also, William H. Becker, "The Impact of America's Becoming Corporate: A Review Essay," *Journal of Policy History* 5, no. 3 (1993), 359, for the contention, similar to my own, that middle managers in oligopolistic industries were "autonomous in how they worked out the directions they received," but that "the environment they worked in was by no means one entirely or fundamentally of their own making." Becker's essay reached me just after this section was complete and I was about to send the manuscript to the press.

6. Munroe Scrapbook; Roy G. Munroe to George E. Lewis, August 18, 1950, Munroe files.

7. Munroe Scrapbook.

Fig. 7. Denver Gas New Business Department, December 2, 1929. By the late 1920s, Roy G. Munroe presided over a large sales organization. Munroe is seated in the first row, eighth from the left, eyes closed. (Courtesy, Public Service Company of Colorado)

later, he remembered that the company had launched a major sales effort in 1923; as for his promotion in 1924 to direct that effort in Denver, he reported, again years later: "I was assigned to head the . . . gas department." More generally, Munroe perceived himself as a successful salesperson, a competent executive, and an employee who took pride in his association with several of the nation's leading utility executives. He had come to understand the importance of growth and change in his city, including an understanding of the city's diverse subcultures and their unceasing movement across Denver's landscape. Consistent with the assumptions of his own gender and time, Munroe had also learned to identify men and women with different responsibilities for creating conditions of cleanliness, comfort, and convenience in the home. As director of a sales campaign aimed at persuading Denver residents to heat their homes and apartments with gas, Munroe emerged as one of the key mediators between goals and specifications issued in New York and the social and economic circumstances that shaped technological tastes in Denver.[8]

8. Ibid.

The largest sales effort of the 1920s in the gas utility industry was directed at heating homes. The idea itself was an old one—since before 1900, gas had been sold for use in ranges that also heated kitchens. Yet distributors had been reluctant to promote gas for house heating, fearing demand would exceed capacity during chilly winter evenings (leading to the expense of installing additional distribution lines to serve only a few momentary peaks). In the absence of additional capacity, however, operators worried about service interruptions and dissatisfied customers, which might jeopardize the franchise. During the mid-1920s, as planning for the home heating campaign was getting under way, Munroe himself was concerned that the "winter peak . . . appeared too steep for further accentuation." On the other hand, executives of the Denver company and their superiors in New York believed that the market for gas stoves and manually operated hot-water heaters was nearly saturated. Consequently, they decided to undertake the risks and expenses associated with heating homes and offices, provided that sales personnel built the gas load in a fashion that kept their industrial, commercial, and space-heating loads in balance.[9]

As a point of departure for the forthcoming sales effort, executives needed cost and load data on which to base the campaign. Munroe's assignment was to secure that data by persuading a limited number of householders to allow company employees to remove coal stoves and heaters and install a furnace that burned manufactured gas. He had started with himself. In 1915, Munroe installed the first gas furnace in Denver in his own home. The cost of heating with manufactured gas, as he later reported, was "prohibitive for the average householder." Cheap coal in the Denver region added to the relative cost disadvantages of heating with manufactured gas that was expensive and low in heat value. Consequently, executives determined that the company would finance the expense of converting coal furnaces to gas, leaving the customer to pay the subsequent gas bills. Few householders participated in that program. In 1918, Munroe secured the first installation of a furnace designed specifically to burn gas, heating hot water for radiators. At a cost to that householder of $1,000 a year for gas, company officials abandoned their efforts until 1922, at which point they were prepared to offer a lower rate. Initially, then,

9. Roy G. Munroe, "Selective Load Building Makes Even Load Curve," *Gas Age Record* 60 (October 8, 1927), 545–546, 550; Munroe Scrapbook.

Munroe's efforts were directed less at sales and more at gauging the potential demand of consumers.[10]

In 1922, senior executives at Denver Gas & Electric launched another program aimed at securing house-heating customers on an experimental basis. Late in September 1922, the company published an advertisement in a local newspaper asking prospective customers for gas house-heat to contact the company. The idea was to identify 100 customers located in different sections of the city. As part of his own preparation for the upcoming promotion, Munroe entered an essay contest sponsored by the company. In April 1923, officials awarded first prize to Munroe's essay, which carried with it a cash award of $200. Munroe's essay consisted of fictional conversations between sales personnel and an assortment of customers, such as a widow with several children, and a prominent and influential attorney. In each case, sales representatives emphasized to incredulous listeners the idea of lower unit rates based on volume consumption. In 1923, as another promotional campaign got under way, company officials, including Munroe, continued to attend to the diversity of their city and to the advantages of mass sales.[11]

Securing those first 100 customers proved a challenge. Munroe and his supervisors turned to the time-tested methods of the Doherty organization. First, they prepared the way with a massive advertising campaign. Second, they organized a large sales force, providing members with training in the details of gas operations and home heating and charging them to canvass every householder in the city. Munroe, generally a perceptive observer of the political economy of sales, described this promotional effort to a gas executive in New York as "the biggest thing which our company has attempted, not only from the gross and net revenue standpoint but also from the public policy viewpoint."[12]

Advertisements for gas-fired home furnaces stressed the obligation of men to shape a healthful environment for members of their families, especially women. Gas heating, the company asserted, offered a level of household amenity unavailable to those heating with coal. Gas

10. Ibid.; Munroe to Lewis, August 18, 1950; Roy G. Munroe, undated interview; T. M. Foulk and T. G. Storey, "An Experiment in Residence Gas Heating at Denver, Colorado," in American Gas Association, *Fifth Annual Convention Proceedings* (New York, c. 1923), pp. 1083–1105.

11. Minutes of house-heating meeting, September 20, 1922; Roy G. Munroe to Henry O. Loebell, January 3, 1923; Roy G. Munroe, "How to Sell the Three Part Rate," *Doherty News* 7 (April 2, 1923), 3–7—all in PSCC.

12. Munroe to Loebell, January 4, 1923.

heating was automatic. It promised additional leisure time with family members, "instant . . . regulat[ion] [of] the heat to suit our changeable climate . . . ," and protection of health, because the heat was regulated against "numerous wintertime coughs and colds." Coal, which represented the major competition for the house-heating market, was decidedly less costly. Rather than discussing price, however, executives emphasized cleanliness and convenience. The virtues of eliminating coal dust and smoke were routinely stressed in company advertisements, as was the "sense of guilt which rests upon all men as they sit in their comfortable downtown offices knowing that their mothers or wives are doing janitor or stoker work at home."[13]

Environmental controls, including the obligation of men to protect women, was only one facet of the march toward progress that executives liked to associate with their products. Officials at the gas and electric company in Denver recognized that residents lacked familiarity with devices such as gas-fired furnaces, which differed in appearance and operation from coal-burning units. As early as 1923, designers of advertising literature at Denver's gas and electric company pointed to replacement during the past two decades of horsecars, kerosene lamps, and manual water pumps by "more efficient methods of public service." Heating with gas was not really unfamiliar; it merely employed "the invisible furnaceman" instead of the hand-stoking that had to be performed in every household heated "by the old fashioned method of solid fuel furnaces." Even the thermostat represented but another step along a familiar path: "It operates with the ease with which you turn a faucet to obtain water, so that it might be termed as 'heat on tap.'" Consequently, the gas furnace was a natural, inevitable, and certainly a "progressive step" in the evolution of

13. PSCC, *The Story of the Invisible Furnaceman, Being a Short Statement of Another Modern Convenience That Will Make Your Home a Healthier and Happier Place in Which to Live* (Denver, 1923). For instructions to Doherty sales managers nationwide emphasizing health, comfort, and convenience, see Henry L. Doherty & Company, "New Business Department Bulletin," January 1924, p. 7, both in PSCC. For an analysis of the genre of advertising that historian Roland Marchand labels "soft focus," see Marchand's *Advertising the American Dream: Making Way for Modernity* (Berkeley and Los Angeles: University of California Press, 1985), pp. 248–254. See also William B. Waits, *The Modern Christmas in America: A Cultural History of Gift Giving* (New York: New York University Press, 1993), pp. 90, 98, 109, for the increasing emphasis (between 1900 and 1940) that advertisers placed on household appliances as gifts for women. On campaigns during the 1920s and 1930s aimed at persuading women in Great Britain to use electricity rather than gas or coal, see Bill Luckin, *Questions of Power: Electricity and Environment in Inter-War Britain* (Manchester and New York: Manchester University Press, 1990), pp. 39–51.

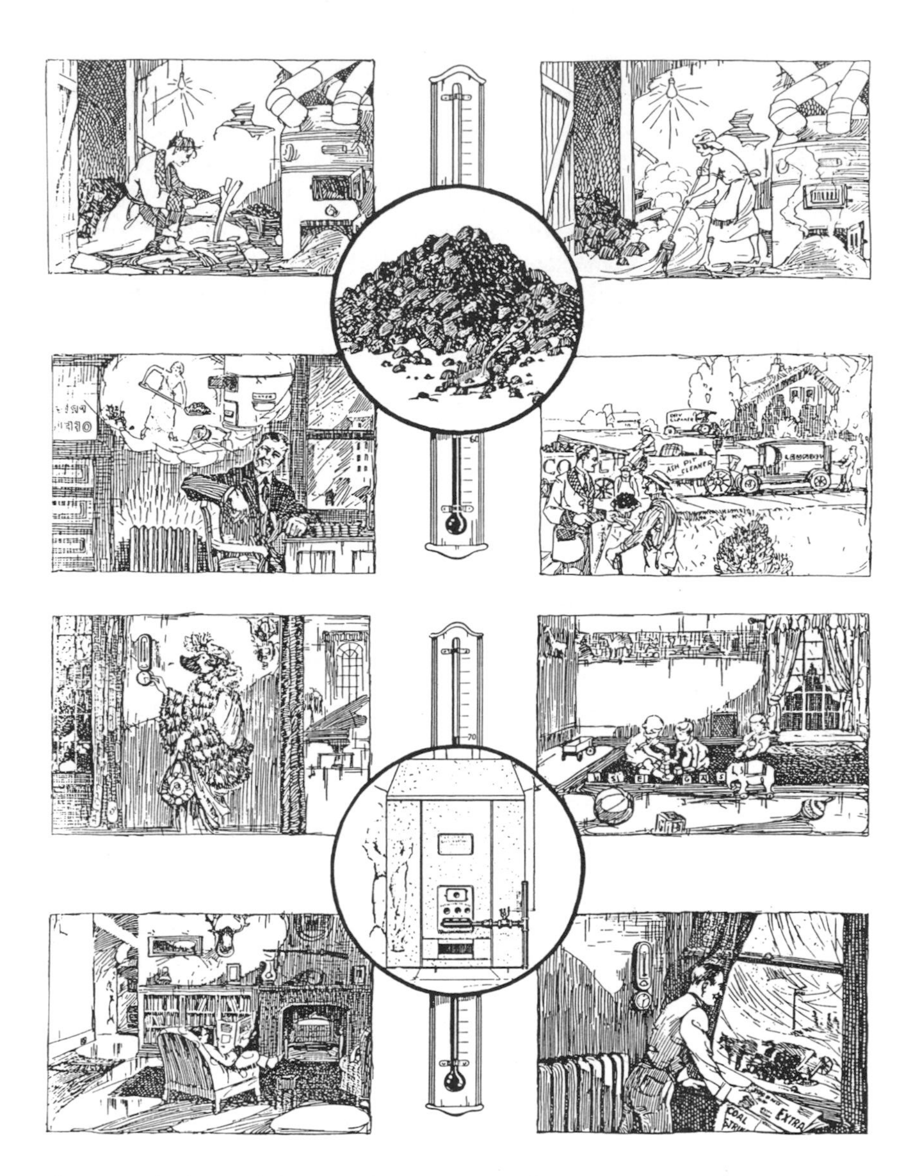

Fig. 8. *The Story of the Invisible Furnaceman* from Denver Gas & Electric, 1923. Promotion for gas and electric appliances often focused on the responsibility of men for the comfort and convenience of women. This ad pointed out the high costs for coal and cleaning that only a wealthy man could afford, for which he received a house temperature of only 60°. With a gas furnace, however, the same man knew that his wife had a warm and clean home. In turn, he could relax in front of a gas heater (in fireplace) and look out the window during a storm. The Invisible Furnaceman (slightly visible at left of furnace) now heated his home to 70°. Others suffered through a coal strike. (Source: Public Service Company of Colorado)

household appliances, much in the same fashion as the "Hoover has overthrown the broom."[14]

In all, Munroe's messages rested on old and powerful ideas such as comfort, convenience, cleanliness, hygiene, and modernity. Yet the gas house-heating campaign also resonated with two other themes, those of the emerging ideal of the companionate marriage alongside the continuing reality of the patriarchal household. After 1900, progressive reformers launched campaigns to protect women and children from danger and abuse; by the 1920s, they had achieved considerable results. Simultaneously, upper-income men were beginning to participate in decorating and other household realms formerly occupied by their wives alone. Munroe and his superiors, then, came quite close to the cultural essence of the matter. A man, comfortable in his office and up-to-date technologically, could no longer permit his wife (or mother) to perform harsh, dirty, unhealthful, and dangerous tasks such as tending a coal furnace.[15]

Munroe personally took charge of preparing the sales force that brought these messages to Denver's householders. According to an account given years later, he spent "months . . . in the hiring of additional sales representatives." In turn, Munroe organized new and existing employees into one of several divisions, including domestic gas, industrial and commercial sales, and house heating. Next, Munroe

14. PSCC, *Invisible Furnaceman.*

15. Nancy Tomes, "The Private Side of Public Health . . ." *Bulletin of the History of Medicine* 64 (Winter 1990), 509–539. For men in the suburbs, see Margaret Marsh, "Suburban Men and Masculine Domesticity, 1870–1915," in Mark C. Carnes and Clyde Griffen, eds., *Meanings for Manhood: Constructions of Masculinity in Victorian America* (Chicago: University of Chicago Press, 1990), pp. 111–127. During the 1980s and early 1990s, the amount of scholarly literature focused on historical dimensions of women, the household, and work has increased rapidly. Rather than citing several monographs dealing with particular examples of efforts to protect women, I continue to find compelling the more broadly framed analysis of Joan Jacobs Brumberg, "Zenanas and Girlless Villages: The Ethnology of American Evangelical Women, 1870–1910," *Journal of American History* 69 (September 1982), 348, 358, 367–369, which reviews ethnographic literature prepared by missionaries in distant and presumably uncivilized lands and the meaning of that corpus of writing for ordinary Americans in terms of the desirability of protecting women and children. For women's magazines and the use of guilt in advertising, see Ruth Schwartz Cowan, *More Work for Mother: The Ironies of Household Technology* . . . (New York: Basic Books, 1983), p. 187; and Bonnie J. Fox, "Selling the Mechanized Household: 70 years of Ads in Ladies' Home Journal," *Gender and Society* 4 (March 1990), 26–27. Finally, see Daniel Horowitz, *The Morality of Spending: Attitudes Toward the Consumer Society in America, 1875–1940* (Baltimore: Johns Hopkins University Press, 1985), pp. 67–108, for an account of the ambivalence of middle-class moralists regarding expenditures (rather than saving) for the products and services that were becoming available during the early decades of the twentieth century.

assigned representatives to sales territories and instructed them to work in conjunction with one another.[16]

As in every Doherty promotion, Munroe insisted that personnel familiarize themselves with the desires of customers. During the period up to about 1910, Doherty had relied on sales representatives such as Munroe to discover the needs of customers and to report them to each other in daily meetings. Inevitably, salespersons had determined that appeals to cleanliness, comfort, convenience, or economy of effort were most effective in securing a sale. In turn, the company published their reports in bulletins for review and study. In the 1920s, Doherty as well as national gas (and electric) organizations began to publish instructions on identifying the customer's preferences in the form of detailed tracts. One instructional program, entitled "Domestic Gas Salesmanship," consisted of specialized units of study such as "Explaining the Appliance" and "Finding Out the Customer's Needs." By 1929, executives of the American Gas Association had begun to codify the sales presentation. An association publication informed sales personnel that a householder faced "the problem of keeping his house healthfully, comfortably, cleanly, and evenly warm at all times, with as little work as possible." A list of seventeen items such as "It does away with the ash, dust, smoke and soot" and "[Gas] means better health for adults and children" were intended to provide "big selling points." In the sale of furnaces and other gas appliances, prescription was replacing the unpredictable results of experience.[17]

The use of gender in sales presentations also assumed a codified form during these years. Before 1910, sales representatives had acted on their own experiences as members of the middle class to formulate arguments deemed appropriate for men and women. Thus, they had extended traditional middle-class gender stereotypes on to these new appliances. During the 1920s, company executives took these same sets of stereotypes for granted, but attached them specifically to the experiences of idealized and invariably wealthy householders. Sales

16. Munroe Scrapbook.

17. American Gas Association, "Building the Gas Load," Unit V, c. 1927; Business Training Corporation under the Supervision of the American Gas Association, "Explaining the Appliance," Unit II (New York, 1929), 44–48 (quotations on pp. 44–45 and 47)—both in PSCC. Finally, see such articles as "Boosting Hot Water Sales with a Cleanliness Appeal," *Gas Age Record* 61 (June 2, 1928), 803–804, in which gas company executives could learn that leaders of the Cleanliness Institute, established by the nation's soap manufacturers, were promoting the sale of soap products on the basis of "beauty, health, character, success, and popularity" and that if cleanliness and gas-fired hot water "can be combined . . . , then the Cleanliness Institute has a real message for the gas man."

personnel learned that these householders, male and female, still held different responsibilities for the maintenance of their large homes. Instructional literature for Munroe's employees depicted men who were concerned about their family's sleep and warmth and about reducing the burden on their wives. But it was women, sales personnel learned, who had to be sold on specific features of appliances. Female pronouns replaced male pronouns as salesmen were told to point out that "her children will be protected against the fluctuating temperatures of the old methods of heating."[18]

However sophisticated the promotional efforts of the 1920s, results in the form of furnaces actually installed between 1922 and 1928 were only modest. In January 1923, Munroe had to report to a gas executive in New York his "regret that it ha[d] not been possible to secure the desired 100 installations before this date." Worse yet, sales personnel had already prepared a list for Munroe's scrutiny of customers who were contemplating removal of their furnaces. By early 1928, following five more years of concentrated effort, company records listed fewer than 400 house-heating installations. "These were the customers who stayed with us," Munroe later reported, "compared with the hundreds who tried gas and later rejected it."[19]

Cost of operation was the major barrier to a more successful sales effort. In January 1923, Munroe's sales personnel reported customer dissatisfaction with high gas bills, often nearly double their costs the year before when using coal. As late as June 20, 1924, Munroe informed a senior executive in Denver: "We have been besieged with requests to take in house-heating equipment and make credit allowances." In reality, costs for gas fell within the estimates Munroe's representatives provided each potential customer, but abstract estimates apparently mattered less than the reality of a high bill, especially when, as it turned out, gas furnaces occasionally failed. Munroe recognized, as did several of the senior engineers in Denver, that gas for house heating was "not competitive with Denver coal except by excessive capitalization of the factor of convenience." Late in 1927, then, most of the nearly 400 householders heating with gas belonged to

18. Business Training Corporation under the Direction and Supervision of the American Gas Association, "Explaining the Appliance," pp. 20–21, 47; Marchand, *Advertising the American Dream,* pp. 248–254, impressions gained from a systematic review of industry literature published during the 1920s and 1930s.

19. Munroe to Loebell, January 4, 1923; Installations That May Be Discontinued on Account of High Bills, Etc., January 1923, in PSCC; Munroe to Lewis, August 18, 1950.

the group Munroe had described in 1923 as "the moderately wealthy class."[20]

Beginning in the summer of 1928, Munroe and his representatives concentrated their training and experience on another gas house-heating campaign, this one featuring the arrival of natural gas. Between June 23 and August 1, 1928, natural gas reached Denver in a pipeline from a field near Amarillo, Texas, a distance of 340 miles. Although many gas suppliers had offered natural gas during the nineteenth and early twentieth centuries to nearby cities such as Kansas City, Denver was among a small number of cities (including San Francisco, Memphis, and Atlanta) that had access to natural gas carried long distances through pipelines. Rates for natural gas in Denver were higher than for manufactured gas, but the natural product contained 2.6 times the heat value, which meant that householders needed less than half as much gas to cook meals or heat houses and apartments. Roy Munroe had participated in the company's plans for natural gas by testing the earliest furnace, by preparing an essay on sales presentations, and by creating a sales force eager to secure the load required to cover the costs of a large investment. He had not participated in tedious negotiations with city officials about rates and service. His assignment was to boost the consumption of gas within a framework arranged by others.[21]

The gas house-heating campaign launched by Munroe in 1928 combined most of the elements that were becoming routine throughout Doherty's large organization. Representatives canvassed each householder and business in the city. Normally, sales personnel worked a territory in teams of two, with one promoting furnaces and the other handling the remaining gas appliances, such as water heaters and stoves. An estimator accompanied these representatives, measuring each house in order to provide a reliable judgment of the quantity and cost of gas required. Because of the importance that

20. Installations That May Be Discontinued . . . ; Munroe to Loebell, January 4, 1923; T. M. Foulk and T. G. Storey, "House Heating—What Denver Found Out About It," *Gas Age Record* 52 (October 6, 1923), 393; Roy G. Munroe to G. B. Buck, June 20, 1924. For the clustering of the earliest householders to heat with gas in high-income neighborhoods such as the Country Club District and Park Hill, see Munroe's copy of House Heating Installations by Domestic Territories, c. late 1927, both in PSCC.

21. For the caloric value of Denver's manufactured and natural-gas supplies and other measures of consumption, see George Wehrle, "Substitution of Natural for Manufactured Gas," *Gas Age Record* 66 (October 25, 1930), 675–680; and Roy G. Munroe to Clare N. Stannard, April 29, 1931, PSCC. F. T. Parks, "Introducing Natural Gas into Denver," *Natural Gas* 9 (August 1928), 3–5, 50, 52, sketches the organization of the pipeline corporation and the introduction of gas in Denver.

Munroe and other executives attached to this campaign, even the estimator merited attention in advertisements, with one item prepared under Munroe's supervision describing that person as an "expert [who] will do the figuring for you." As in every promotion, moreover, attractive terms were made available, ranging as long as fifty-four or more months to pay and only 10 percent down for a combined house-heating and hot-water installation. Local utility operators, like automobile manufacturers, were by then financing their own sales. Billboards, newspaper advertisements, and direct-mail solicitations—by the thousands—supplemented the sales efforts, and Munroe also supervised the offering of prizes and bonuses to successful sales representatives. Because Munroe and other officials stressed controlled load-building rather than profits on appliance sales, the price of appliances remained low, and often the Gas New Business Department operated at a loss.[22]

Results of the gas-fired furnace campaign initially exceeded expectations and then fell off. Early in 1928, a point at which fewer than 400 householders heated with manufactured gas, a company official projected more than 6,600 natural-gas installations within two years. By November 1929, only fifteen months after the arrival of gas from Amarillo, nearly 6,350 households were on company mains. Unemployment and falling coal prices slowed the rapid upswing in heating installations. By May 1932, however, nearly 10,200 households purchased gas for heating. During the next two years, the company and local dealers continued to sell gas furnaces and boilers, yet discontinuations occurred at a faster rate, due in part to nonpayment of bills, vacancies, and cheaper coal. According to a memo prepared at the gas company for internal distribution, the average annual cost of heating a house with coal was $80, compared with $120 for gas, a difference of 50 percent. Most likely, coal operators and dealers accelerated the decline in the number of gas furnaces by marketing automatic stokers and publishing advertisements asking "Coal with Safety and Economy or Gas with Danger and Expense?" According to Munroe, then, the company faced a "buyers' strike" brought on by falling coal prices, reduced incomes, and "prejudice created by newspaper attacks."[23]

22. "Our House Heating Experts Will Do the Figuring for You" (advertisement), c. November 1928, one of a large number of advertisements located in the PSCC Records.

23. G. B. Buck, "Gas Division Report and Natural Gas Expectancies," March 17, 1928; Roy G. Munroe to V. L. Board, July 26, 1933; Roy G. Munroe to G. B. Buck, August 28, 1933, August 30, 1933; Househeating Data, February 13, 1931—all in PSCC Records. "Coal with Safety and Economy or Gas with Danger and Great Expense"

Fig. 9. "This Year Give a Useful Gift," which in this case was a space heater, December 1931. The idea in every sales promotion was to attach new appliances to familiar ideas and practices. (Courtesy, Public Service Company of Colorado)

By the early 1930s, Munroe and his assistants recognized that the gas house-heating campaign had run its course. In December 1931, one of Munroe's superintendents reported to him that "gas has been sold far beyond the true economics of the situation." As it stood, he added, "all intangibles have been highly capitalized including firing,

(advertisement), *Denver Post,* September 23, 1929; Carl B. Wyckoff, "Holding House-Heating Load for Today and Tomorrow," *Gas Age Record* 72 (July 8, 1933), 31–34, 42.

labor, smoke nuisance, [and] ash removal." Not even a price reduction of 13 percent for gas authorized in August 1934 attracted many more customers to gas heating. In 1934, Munroe calculated that 16 percent of those residing in single and double household buildings in Denver (with company mains nearby) were heating with gas. By 1936, following an intensive effort to return old customers to gas heating, the number crept up to 20 percent, the point at which it stayed until the conclusion of the decade.[24]

Conclusion of the gas house-heating campaign also led company officials to reassign Munroe. On June 5, 1935, he learned that his post as Gas New Business Manager was abolished and that he was to organize another sales activity. As part of an industry-wide trend, gas executives planned to eliminate door-to-door sales work in favor of coordinating appliance sales through local retailers. In November 1936, Munroe prepared a lengthy memo detailing the advantages of door-to-door solicitation, particularly in light of "sales resistance, . . . which requires so much specialized semi-technical knowledge of both sales methods and equipment." But the decision had been made. In January 1938, Munroe became manager of electric appliance sales, including the showroom at company headquarters and other units that he later described to a gas executive as "some odds and ends." By the late 1940s, Munroe headed the Dealers' and Builders' Department, "assisting dealers and plumbers to do the merchandising that we formerly had to do for introductory reasons."[25]

Resentment at what must have seemed a demotion was not in Roy Munroe's nature. Although most of the substance of Munroe's correspondence dealt with the details of shaping a sales organization and boosting demand for appliances, occasionally he and his correspondents allowed themselves a moment of reflection. In the first place, Munroe had embraced the hierarchical nature of relationships within the corporation. In an interview with his granddaughter many years after retirement, Munroe described one of his superiors for the period of the late 1920s and early 1930s as "a most astute executive, and a wonderful boss and friend." Two other executives, as he re-

24. C. W. Gale to R. G. Munroe, December 31, 1931; PSCC, "The Gas Pilot," August 25, 1934; PSCC, Denver District, "Proposed Reorganization of Denver Residential Gas Sales Department," November 1, 1936; Roy G. Munroe to G. B. Buck, July 17, 1939—all in PSCC. Munroe's calculations of the percentages of Denver householders burning gas for heating did not include those residing in apartment houses.

25. Munroe Scrapbook; PSCC, Denver District, "Proposed Reorganization"; Roy G. Munroe to Henry O. Loebell, October 21, 1940; Roy G. Munroe to B. P. Montagriff, December 2, 1949—all in PSCC Records.

called, had "befriended me in a personal way, praised my accomplishments and boosted me along in a business way." In the second place, Munroe also endorsed the company's broader goals of increasing the demand for gas and the prospect that burning gas as a replacement for coal would help clean the built and out-of-doors environments. During the late 1930s, as his accomplishments in promoting gas for water and house heating received attention from the editors of gas and engineering periodicals, Munroe boasted of Denver having "the highest percentage of saturation of any city of comparable population." Whether or not this was actually the case, he nevertheless wrote an executive of the gas company in Dallas in March 1939 that "all gas men should set as their goal . . . [a] skyline without a single coal-smoke blemish."[26] J. C. Nichols, builder of the beautiful Country Club District in Kansas City, entertained identical goals. Nichols operated with a great deal more latitude.

In 1880, Jesse Clyde Nichols was born on a farm near Olathe, Kansas, a small town several miles southwest of Kansas City. His adolescent and collegiate years were those of a bright, personable, and articulate male born to a moderately successful family. Nichols was an excellent student, and popular as well. At the nearby University of Kansas, he majored in economics, earned election to Phi Beta Kappa, and served as president of his class.[27]

During the summer of 1900, before his senior year at the university, Nichols toured Europe by bicycle. He and a friend, Wilkie Clock, a divinity student at Harvard University, worked their way across the Atlantic as hands on a cattleboat. While in Europe, Nichols prepared several descriptive essays for his hometown newspaper, the *Olathe Mirror.* These articles are not only a chronicle of what he heard and saw, but also a record of the elements of housing, transport, land-use patterns, aesthetics, and social class that Nichols deemed most important as he began to contemplate his own future. In London, for instance, they "went to bed early and slept to the music of the numer-

26. Munroe Scrapbook; Roy G. Munroe to L. C. Kerrick, January 9, 1939; Roy G. Munroe to C. K. Patton, March 17, 1939. See also C. W. Gale to Roy G. Munroe, August 31, 1931, for the observation "Pope says I am the best friend he has in the world, I think you are the best friend I have in the world and you say Mr. Stannard is the best friend you have in the world"—all in PSCC Records.

27. A. Theodore Brown and Lyle W. Dorsett, *K.C.: A History of Kansas City, Missouri* (Boulder, Colo.: Pruitt Publishing Company, 1978), pp. 169–171; William S. Worley, *J. C. Nichols and the Shaping of Kansas City: Innovation in Planned Residential Communities* (Columbia: University of Missouri Press, 1990), pp. 63–64.

ous chapel bells"; around Liverpool, Nichols found that "every home is surrounded by parks or beautiful lawns and flower gardens" and that "the roads are paved and generally smooth as glass." Later, he found Paris a city of "wide streets clean and straight," but lacking a transit system adequate for "rushing Americans."[28] In short, Nichols took the view, a minority one in the American tradition, that business and beauty were compatible principles.

In 1901, Nichols began a year of postgraduate study at Harvard University. One professor in particular, economist Oliver M. W. Sprague, impressed Nichols with the observation that industry was moving away from the Northeast and toward the South and West. The trick, then, was to position real-estate investments at the edge of the line of settlement.[29]

Nichols graduated from Harvard and moved to Texas, which was in the middle of a land boom. His idea was to purchase large tracts of land and develop agricultural communities. Because Nichols selected a site ahead of the line of settlement, he believed he would be in a position to take advantage of a vast increase in demand soon to arrive. Although the project itself was a "dismal failure," according to historian Richard Longstreth, Nichols's experience in Texas encouraged "a taste for Spanish colonial architecture in the process."[30]

In 1903, Nichols returned to Kansas City and entered the home construction business. Loans from his father, several farmers in Olathe, and W. T. and Frank Reed, local attorneys, financed this venture, which did business as Reed, Nichols & Company. Nichols purchased 10 acres of land near the industrial district in Kansas City, Kansas, and directed construction of 100 homes for workers. Prices were modest—less than $1,000 per house. During a two-year period, the net profit on this undertaking amounted to the substantial sum of $18,000, sufficient to begin his next project, the Country Club District in Kansas City, Missouri.[31]

In 1905, Nichols and his associates purchased a tract of land on Fifty-first Street outside the city limits that resembled country farm-

28. *Olathe Mirror,* August 9, August 16, and August 23, 1900, all in the J. C. Nichols Papers, UMKC (cited hereafter as Nichols Papers); Richard Longstreth, "J. C. Nichols, the Country Club Plaza, and Notions of Modernity," *Harvard Architectural Review* 5 (1986), 123, a copy of which Howard Gillette Jr. was kind enough to send to me.

29. Worley, *Nichols and the Shaping of Kansas City,* p. 64.

30. Ibid.; Longstreth, "Nichols, the Country Club Plaza, and Notions of Modernity," p. 123.

31. "J. C. Nichols Builds Again," *Architectural Forum* 61 (October 1934), 302–303; Brown and Dorsett, *K.C.,* p. 172.

land rather than a "country club district." Nichols applied that title because of the area's proximity to Kansas City's first country club. But the streets were not paved and improvements such as water and sewer service were not available. A feeding lot for hogs, a farm, and a smoke-belching brickyard were located nearby. Nichols and another salesperson, John C. Taylor, picked up prospects in a horse and buggy at the end of the trolley line and brought them to the subdivision, a trip of about one mile. Nichols later admitted that he could only promise that "the city is bound to move out this way, it's the logical place for homes." Nichols also acted quickly to ensure that buyers continued to appear, arranging for electric trolley service to the district in 1907. Two years later, the city limits were extended to Seventy-ninth Street, which guaranteed access to the full range of urban services.[32]

Yet Nichols chose not to wait for the arrival of public officials and their services for achieving environmental improvements. "Spring days make you yearn for a home," he advertised in April 1906, and "health demands freedom from coal smoke and dust." His recommended solution was to "deed your wife a nice new 5 or 6-room cottage, [each with] flowers, grass, trees, city water, [and] electric light." The following month, Nichols advertised "sidewalks already in"; in October, he was offering houses with a "sink in kitchen [and] plum[bing] for bath and furnace." By 1909, water, electricity, and paved roads, as well as the trolley, were available.[33]

Modern technologies and expensive homes were made to coalesce with traditional landscape themes, the sort Nichols had seen in Europe and that were appearing regularly in elite suburbs in the United States. By 1908, he was advertising "little stone fences, shelters, parkways, rustic bridges, and other pleasing features." In the Country Club District, Nichols reported, "birds sing and air is pure." During the next few years, Nichols hired an expert from Massachusetts "to lecture on birds" and even awarded "prizes to the children" in order to "get the whole people thinking about birds." Flowers and flower

32. Ibid., pp. 172–173; "Nichols Builds Again," p. 303; Jesse C. Nichols, "When You Buy a Home Site," *Good Housekeeping* 76 (February 1923), 38–39, 172–176, a semi-autobiographical essay generously called to my attention by William Worley. For Nichols as one of several builders who were helping to determine the design, demography, zoning, and financing of upper-income suburbs, see Marc A. Weiss, *The Rise of the Community Builders: The American Real Estate Industry and Urban Land Planning* (New York: Columbia University Press, 1987).

33. *Kansas City Star,* various dates, as cited in William S. Worley, "J. C. Nichols and the Country Club District: Origins and Development of Community Consciousness" (seminar paper, Department of History, University of Kansas, May 1977), which the author prepared for a seminar directed by John G. Clark and me.

gardens also occupied the attention of Nichols and his buyers. After 1910, purchase of a home in the Country Club District allowed householders to merge ornithological and botanical study with the comforts of modern technological systems.[34]

During the 1920s and 1930s, Nichols continued to emphasize birds, parks, trees, flowers, and golf courses as part of a strategic setting for homes that were increasingly filled with modern appliances. Nichols and his employees believed that domestic appliances boosted home sales. According to Edward W. Tanner, Nichols's top architect, homes for the well-to-do had once been designed to last 200 years, but prospective buyers were no longer interested, he believed, in making a long-term investment in ornamentation and construction. During the interwar decades, he designed homes featuring "automatic heating plants, wall and ceiling insulation, [and] complete electric installations." The "public," as he characterized the limited portion of home buyers who dealt with Nichols's firm, wanted to forsake the "thoroughbred two-hundred-year house" in favor of "fine mechanical installations."[35]

Space and water heating equipment were among several expressions of the triumph of mechanics in shaping the built environment. The exhibition house for 1925 included an oil-fired water heater described as "instantaneous," a popular term in the trade. In a house built in 1926, a Nichols publication pointed out that "a specially designed gas boiler is a basement feature." In 1926, Nichols advertised an all-gas house that featured a gas-fired furnace and hot-water heater, exactly the type of equipment Roy Munroe was getting ready to merchandise in large numbers in Denver. As purchasers of homes in the Country Club District installed gas rather than coal-burning

34. As cited in Worley, "Nichols and the Country Club District." For the origins of elite suburbs, see Michael H. Ebner, *Creating Chicago's North Shore: A Suburban History* (Chicago: University of Chicago Press, 1988); and Carol A. O'Connor, *A Sort of Utopia: Scarsdale, 1891–1981* (Albany, N.Y.: SUNY Press, 1983). For antimodernism as a component of the aesthetic sensibilities of householders in elite suburbs, see T. J. Jackson Lears, *No Place of Grace: Antimodernism and the Transformation of American Culture* (New York: Pantheon Books, 1981); and Mary Corbin Sies, "The City Transformed: Nature, Technology, and the Suburban Ideal, 1877–1917," *Journal of Urban History* 14 (November 1987), 81–111.

35. "Problems in Designing Houses for Today's Buyers," *National Real Estate Journal* 40 (February 1939), 66. Tanner contended that he had shifted his emphasis in favor of the 100-year house after 1927. In reality, he was focusing on one comparison and overlooking countless homes sold at comparatively modest prices throughout the firm's history. See photographs and descriptions of houses constructed during the early to mid-1920s in monthly issues of the *Country Club District (CCD) Bulletin,* located in Nichols Papers.

Fig. 10. J. C. Nichols, developer of the Country Club District and Country Club Plaza, Kansas City, Missouri, no date. (Courtesy, Western Historical Manuscript Collection, University of Missouri at Kansas City)

Fig. 11. Exterior and interior of a modestly sized and priced home (by the standards of the Country Club District) built by J. C. Nichols during the early 1920s. Hidden beneath the lawn and inside walls and floors were the pipes and wires that furnished gas and electric service to a number of lights and appliances that only a decade before would have been unimaginable (or unavailable) for all but the wealthiest householders. Yet Nichols, like most builders, wrapped the house's infrastructure in a conventional design. In short, the future came packaged as the present or even the past. (Courtesy, Western Historical Manuscript Collection, University of Missouri at Kansas City)

equipment, a Nichols publication was able to boast that "there is less smoke in the district."[36]

Electric and gas appliances for the kitchen were the central elements in Nichols's efforts to appeal to women who were seeking comfort, convenience, and environmental control. In his description of a house constructed during 1926, Nichols boasted of "a ventilating system in the kitchen that will remove every cookery odor." In another house constructed that year, Nichols joined with the local gas company to build a colonial style residence featuring a Roper gas stove in the kitchen. By 1925, Nichols reported that electric lights had been arranged "to keep the kitchen free from shadows." Consistent with architect Tanner's metaphor of the kitchen as a laboratory, during the mid-1920s a Nichols publicist described his kitchen appliances and their arrangement as "scientific." In this instance, a scientific kitchen was "a model for cleanliness and convenience."[37]

Nichols's interests became even more pronounced in the construction of his Country Club Plaza, the first large shopping center built outside the central business district. Construction began in 1922, and three shops opened early the next year. By 1926, nearly 100 businesses were operating on the plaza. Several factors led to its remarkable popularity. In the first place, Nichols situated the plaza near homes and apartments in the Country Club District and at the intersection of major highways. The plaza was accessible to higher-income residents of the Kansas City region, with free parking when they arrived. Kenneth Jackson points out that Nichols's policy of renting space on the second floor of his buildings to dentists, physicians, and attorneys "help[ed] stimulate a constant flow of well-heeled visitors." Once at the plaza, Nichols made shopping easy. Stores were grouped in terms of "scientific principles," a favorite Nichols term that was popular with the literate public, encouraging a flow of customers be-

36. "1925 Exhibition Furnished Home Soon to Open," *CCD Bulletin* 5 (November 1925), 3; "405 Dartmouth Road—Is It the Most Efficient Exhibition Home Yet Developed in Kansas City?" *CCD Bulletin* 7 (October 1926), 2; "Gas Equipment Throughout New Nichols Home," *Gas Service* 2 (September 1, 1926), 3; "Exhibition Home in Greenway Fields of Country Club District," *CCD Bulletin* 5 (October 1924), 2; "There Is Less Smoke in the District," *CCD Bulletin* 5 (February 1925), 2. See also David E. Nye, *Electrifying America* (Cambridge: MIT Press, 1990), p. 266.

37. "Problems in Designing Houses for Today's Buyers," p. 68; and "High Standards Influence Commercial Home Building in Kansas City" —*CCD Bulletin* 7 (July 1926), 6; "1925 Exhibition Furnished Home Soon to Open," p. 3; "Gas Equipment Throughout New Nichols Home," p. 3. For the ritual invocation of science in everyday discourse, see my "Science as an Idiom in the Domain of Technology," *Science and Technology Studies* 5 (Spring 1987), 3–11.

tween stores—"a beauty parlor next to a dry good shop instead of a hardware store or grocery." Nichols also maintained architectural continuity, selecting another period theme, this time Spanish Renaissance. The image was that of the traditional overwhelming modernity, a curious design idea that was emblematic of a modern, upper-income neighborhood.[38]

Shop interiors eschewed period architecture in favor of modern technologies and environmental control. The emphasis in designing Wolferman's grocery store, a Nichols account reported, was on "sanitation and service." Wolferman's owners installed refrigerators, fifteen in all, along with exhaust equipment in each department. "No danger of your food being exposed to the elements," declared a Nichols publicist. The battle against the elements being waged at Wolferman's anticipated by several years the early promotions of electric company executives who advertised domestic refrigerators as part of an effort to "impress on the American public . . . the cause of food decay and bacteria growth."[39]

In a Nichols project, moreover, good taste and personal service were never far from technology, hygiene, and the roles assigned to women (and men). The idea at Wolferman's was "to cater to the aesthetic tastes of patrons," reversing the notion that "eggs are eggs" and that buying food is a "prosaic business." Wolferman's employees "radiate[d] pride"—particularly, one assumes, as they handled countless requests from housewives who shopped by telephone and had orders delivered in time to prepare meals that evening. Nichols referred to such a shopper (probably one of the majority even in the fashionable Country Club District) as "the woman who does her own house work." In both literal and symbolic fashion, such stores as Wolferman's on the Country Club Plaza connected aspects of the lives of well-to-do women, particularly shopping, as both an exciting experience and a way of showing devotion to domestic responsibilities.[40]

38. Jackson, *Crabgrass Frontier,* p. 258; Longstreth, "Nichols, the Country Club Plaza, and Notions of Modernity," pp. 129–131; *Country Club Plaza: 46th Street Terrace to Ward Parkway Mill Creek Parkway Westward to Penn* (Kansas City, Mo., 1926), p. 7, in Snyder Collection; Lears, *No Place of Grace,* p. 301; Worley, *Nichols and the Shaping of Kansas City,* pp. 248–259.

39. "Wolferman's Country Club Plaza Store a Masterpiece," *CCD Bulletin* 5 (March 1924), 3; "Nation-wide Food Preservation Campaign to Be Promoted by N.E.L.A. with Other Associations Cooperating," *NELA Bulletin* 16 (June 1929), 366.

40. "Wolferman's A Masterpiece," p. 3; "Phone and Auto Dictate New Design in Western Grocery," *Architectural Record Combined with American Architect and Architecture* 83 (March 1938), 61; "High Standards Influence Commercial Home Building in Kansas City," *CCD Bulletin* 7 (July 1926), 6. The scholarly literature on women, shopping, and

Fig. 12. Kitchen for the "Little Magic House," another model home built by J. C. Nichols, 1939. Increasingly, American kitchens followed the layout of a factory, with production beginning in the refrigerator (not shown) and continuing across the sink, through the mixer, to the electric stove, and then out to the dining room. Perhaps the "woman of the house" telephoned Wolferman's Grocery Store for the "raw materials" of household production. (Courtesy, Western Historical Manuscript Collection, University of Missouri at Kansas City)

In 1930, dull Depression days reached Kansas City and the Country Club District. For several years, Nichols's firm lost money, but diversified holdings and the high incomes of residents restored profitability. In 1934, as Nichols prepared to spend $360,000 on home construction, *Architectural Forum* proclaimed "J. C. Nichols Builds Again." By the late 1930s, Nichols was constructing subdivisions in Johnson

---

changing definitions of domestic responsibilities is large and substantial; for examples, see Cowan, *More Work for Mother,* 172–181; William R. Leach, "Transformations in a Culture of Consumption: Women and Department Stores, 1890–1925," *Journal of American History* 71 (September 1984), 319–342; and Susan Porter Benson, *Counter Cultures: Saleswomen, Managers, and Customers in American Department Stores, 1890–1940* (Urbana: University of Illinois Press, 1986), pp. 4–6.

Fig. 13. Wolferman's Grocery Store, c. 1939. Similar to homes in the Country Club District, Wolferman's exterior was conventional, including the Spanish tile roof preferred by J. C. Nichols. Yet Wolferman's management emphasized new refrigerators to cool produce, an idea that was increasingly important to members of the first generation of Americans who had been educated in the relationships between hygiene, public health, and gas and electric technologies. (Courtesy, Western Historical Manuscript Collection, University of Missouri at Kansas City)

County, Kansas, across the state line but adjoining the original project launched in 1905. Nichols also continued to serve on numerous boards and commissions, most having to do with civic philanthropy, art, and city planning. In 1939, the *National Real Estate Journal* devoted an issue to Nichols and his firm and projects. In 1940, the Metro Club of Kansas City elected him Man of the Year, signaling the appreciation of political and business leaders for his version of a public and private partnership that emphasized controlled growth and creation of a built environment that appealed to persons like themselves.[41]

41. Brown and Dorsett, *K.C.*, p. 178; "J. C. Nichols Builds Again," *Architectural Forum* 61 (October 1934), 302–303; *National Real Estate Journal* 40 (February 1939).

Although Nichols remained a prominent figure for another decade, he had already introduced and institutionalized most of his innovations in design, sales, technology, and land-use planning. After 1930, whether dealing with local politicians, executives of the city's gas and electric companies, or builders, Nichols was content to return to older themes. Newer technologies such as air conditioning appeared in Nichols's buildings, but their installation simply cohered with the ideas of a builder interested in renting to higher-income tenants seeking prestige and environmental control. During 1940 and 1941, then, Nichols's business correspondence focused on prosaic and traditional matters such as gas and electric rates for cooling new buildings, on the widening of streets to "enable people in automobiles to pass through the Country Club Plaza," and on construction of a new swimming pool for the Mission Hills Club that would be "comparable to the Country Club and help maintain our prestige and standing."[42]

Even in dealings with members of his own family, Nichols returned to older ideas about tastes and technologies. The interior arrangements of a home still had to foster comfort, especially for women. As World War II was ending, Nichols wanted to purchase a house for his daughter-in-law and son, who was due to be discharged from the army. "This house," he wrote to his son, "has 3 bedrooms, two sleeping porches on the second floor, servants quarters and a large recreation room on the third floor." Presumably, Nichols's daughter-in-law would handle domestic chores, such as telephoning her grocery order to Wolferman's. Other domestic labor appeared to be out of the question: Nichols added, "Mother thinks you should have a house with servants quarters as when you come home you will be so busy you will not want to wash dishes and clean floors." Throughout his career as a community-builder, J. C. Nichols applied trustworthy concepts for tying together beauty, hygiene, and modern technical systems in shaping the built and out-of-doors environments.[43]

J. C. Nichols was seventy years old when he died in February 1950. Roy G. Munroe, several years younger than Nichols, continued as a

42. J. C. Nichols to John Arthur, March 8, 1940; J. C. Nichols to Board of Park Commissioners, July 7, 1939; J. C. Nichols to James E. Nugent, July 14, 1939; J. C. Nichols to O. G. Bitler, March 21, 1940—all in Nichols Papers. See also J. C. Nichols, "Mistakes We Have Made in Shopping Center Development," *Technical Bulletin* (Urban Land Institute) 4 (August 1945), 2–15, which deals with such matters as street layout, traffic flow, and utilities.

43. J. C. Nichols to Miller Nichols, September 10, 1945. See also Nichols's correspondence with his son for July 16, July 20, July 25, and August 16, 1945—all in Nichols Papers.

manager of sales at the Public Service Company of Colorado until 1954, and then settled into an active retirement as an officer of the Rocky Mountain Gas Association. Nichols had achieved national prominence, but Munroe, a mid-level employee of a large corporation, went unrecognized even in Denver. Similar to their counterparts in every city, however, home builders and electric and gas executives complemented one another in shaping ecologies of technology. In brief, Munroe and Nichols, as sales personnel, helped determine the technological experiences of residents of their cities' vastly different neighborhoods, and the meanings attached to those experiences.

For the period up to 1935, the uneven distribution of gas and electric appliances across the landscapes of Kansas City and Denver is striking. Kenneth Jackson ascribes this uneven outcome to a process he calls stratified diffusion. The numbers show clearly that appliances were clustered in neighborhoods with the highest household incomes. In an area about a mile and a half from Denver's central business district, 60 percent of the primary wage earners in 1933 were unemployed, and a large number were on relief. Around two-thirds enjoyed electric lighting, but fewer than half had gas service. A majority of those Denverites cooked and heated with coal or wood rather than with one of Roy Munroe's new gas-fired furnaces. Only a few miles to the southeast, in Denver's Country Club District, electric and gas service was virtually universal (greater than 99 percent). A list compiled by Munroe in mid-1934 of householders heating with gas showed nearly 250 buildings clustered on only five streets immediately north of the prestigious Denver Country Club. There was less smoke in Denver's fashionable districts too.[44]

As suburban growth resumed after 1935, the process of stratified diffusion meant simply that the newest appliances showed up in the largest numbers at greater and greater distances from downtown. In the summer of 1936, executives of the Kansas City Gas Company installed the first gas-fired air-conditioning unit in the city at their district office and showroom located in J. C. Nichols's Country Club Plaza. In 1938, reported an editor of a gas-industry journal stationed

44. Jackson, *Crabgrass Frontier,* p. 11; F. L. Carmichael, "Employment and Earnings of Heads of Families in Denver," *University of Denver Reports* 10 (September 1934), 6; *Gas Pilot* (August 25, 1934), in PSCC Records. For a similar pattern of gas and electric distribution and utilization in Kansas City, see Missouri Era Non-Manual Works Program in Collaboration with the City Plan Commission of Kansas City of the Jackson County Emergency Relief Committee as Authorized by the Missouri Relief and Reconstruction Commission, *Report of the Housing Survey of Kansas City, Missouri, 1934–1935* (c. 1938), pp. 25–30, plates I, XV.

in Kansas City, each of eighty-five houses J. C. Nichols constructed in a new subdivision featured gas-fired forced-air furnaces and gas-fired hot-water heaters—"and practically all of them use gas for cooking." Nearer downtown, whether in Kansas City or Denver, electricity lit most homes and a majority cooked with gas, but few had central heating. During an earlier period, an observer had described the suburbs that were beginning to surround the nation's cities as "the cool green rim," suggesting the cooling effects of breezes, trees, and open spaces. By the late 1930s, such persons as Nichols and Munroe had taught wealthy residents to define the cool green rim as a built environment that included mechanical refrigeration, forced-air heating and ventilating, and perhaps air conditioning.[45]

Not only had the well-to-do followed the recommendations of J. C. Nichols and Roy Munroe in attending to their own comfort and convenience, but they had attached to their environment and to the appliances that made it possible a set of literal and symbolic meanings that Nichols and Munroe had only suggested. Historian Joan Wallach Scott reminds us that "without meaning, there is no experience [and] without processes of signification, there is no meaning." Signification according to gender, contends Scott, brought meaning to household arrangements, including the appropriate roles of men and women. In turn, Scott's emphasis on gender and the historical division of household labor serves as an introduction to Ruth Schwartz Cowan's equally valuable constructs for making sense of the experiences of men and women in purchasing and using gas and electric appliances in diverse urban contexts.[46]

Cowan divides housewives into "those who were struggling to make ends meet" and others "who lived comfortably." Among women located in the higher-income groups, Cowan finds a curious anomaly. These women purchased household appliances such as vacuum clean-

45. Charles D. Greason, "Gas Captures New Kansas Homes," *Gas Age* 83 (March 30, 1939), 32; Charles D. Taft, "Gas Company's Office in Kansas City Is Air Conditioned," *Gas Age Record* 79 (May 13, 1937), 76, 84; Ernest Manheim in cooperation with Iona Rowland, with the aid of Dr. Worth M. Tippy, *Kansas City and Its Neighborhoods: Facts and Figures* (Kansas City, Mo.: Kansas City Council of Churches and the Department of Sociology of the University of Kansas City, 1943), pp. 47, 50, 56, 59; Denver Bureau of Public Welfare, *The Denver Relief Study: A Study of 304 General Relief Cases Known to the Denver Bureau of Public Welfare on January 15, 1940* (Denver, 1940), p. 52; the observer's description of the urban periphery as the "cool-green-rim" is cited by Richard C. Wade in his "Foreword" to Zane L. Miller, *Boss Cox's Cincinnati: Urban Politics in the Progressive Era* (New York: Oxford University Press, 1968), p. viii.

46. Joan Wallach Scott, *Gender and the Politics of History* (New York: Columbia University Press, 1988), p. 38.

ers and washing machines, substituting machinery and their own labor for expensive helpers. As early as 1924, Nichols recognized this very phenomenon, advertising a "conveniently arranged kitchen" as the answer to "the annoyance of the servant problem." Apart from the wealthiest women, most of the wives in the Country Club District expected to spend their weekdays in a cycle of cleaning, cooking, caring for children, and doing laundry, a routine that was interrupted occasionally by telephone conversations, deliveries from Wolferman's, and "a bridge party at night." Located as they were in neighborhoods in which fewer households burned coal for heating and cooking, these women, through their own efforts, might aspire to an antiseptic household and a family whose health was greatly improved. Thus, reports Cowan, began a process in which, as well-off women turned to machines seeking to improve their families' hygiene and comfort, they also furthered what she labels their own incipient "proletarianization." Subsequently, reports Cowan, this household, with its emphasis on the glory and satisfaction to be derived from housework and childcare, nourished the feminine mystique.[47]

Among working-class women, the same appliances, when they could be purchased or included in a rental unit, facilitated a marked improvement in the health of their families and the physical conditions of their homes. By the late 1930s, the gas stove and a mechanical refrigerator or old-fashioned but usable icebox had achieved widespread diffusion in Kansas City and Denver and an important place in the routines of lower-income households. In 1941, an observer of the housing scene in Denver found that "dwelling units provided with inexpensive equipment for heating, cooking, and refrigeration are more completely occupied than are those more expensively equipped." Women in these households, reports Cowan, perceived gas and electric appliances as tools that enhanced their ability to im-

47. Cowan, *More Work for Mother*, pp. 172–191, and quotations on pp. 154, 160, 180; and Ruth Schwartz Cowan, "Two Washes in the Morning and a Bridge Party at Night: The American Housewife Between the Wars," *Women's Studies* 3 (1976), 147–172; "The Edgevale Road Houses in Armour Hills," *CCD Bulletin* 5 (October 1924), 4. For women's appropriation of the telephone as a device for widening, deepening, and extending relationships and thus as an antimodern device, see Claude S. Fischer, *America Calling: A Social History of the Telephone to 1940* (Berkeley and Los Angeles: University of California Press, 1992), pp. 233–254. For an introduction to the trade literature dealing with an evolving relationship between household appliances and continuing efforts to achieve a hygienic environment, see "Bacterial Rays Eliminate Familiar Health Hazard," *Architectural Record Combined with American Architect and Architecture* 85 (April 1939), 73; and Arthur Cecil Stern, "Control of Air-Borne Bacteria," *Architectural Record* 86 (July 1939), 73.

prove the health of their families and the conditions of their homes. Appliances meant less dust, cleaner sheets and clothing, greater variety in meals, hot water, a warmer house during the winter, and a better-ventilated house during the summer. Among these women, then, the matter was less that of an inchoate mystique surrounding housework or the early stages of proletarianization; appliances were more a matter of the nitty-gritty of enhancing household production, health, and welfare.[48]

Men played different roles in the diffusion of household appliances. Consequently, those appliances held different meanings for them. Men perceived appliances as another arena in which they had to provide for women and children. The number and quality of appliances that men provided varied considerably according to income, educational background, and the type of relationship they had developed with their wives. Late in 1939, social workers visiting the poorest householders located near downtown Denver reported inadequate diets, cold meals, cold rooms, and a resort to the theft of food and coal. Men residing in this area expressed hostility, fear, and rage about their situations. Even if these men lacked a sophisticated knowledge of Munroe's newest appliances, they surely understood what society required of them.[49]

For men who owned homes in the Country Club Districts of Kansas City and Denver, the arguments of salespersons such as Munroe and Nichols resonated with changes they were experiencing with respect to their colleagues and wives. Portions of the lives of these men were undergoing rapid and often dangerous alterations. Although the men who lived in these Country Club Districts enjoyed impressive titles, membership in prestigious clubs, and substantial salaries, they also knew at first hand the volatility of markets and the risks associated with the management of goods, machines, capital, and other men. A modern home built by Nichols, or one like it, retained in a symbolic fashion remnants of an older patriarchy. With its period architecture and Nichols's reiteration of its proximity to nature, homes of this type preserved aspects of the traditional, offering a perception of psychological security and stability in the face of smoke, noise, dirt, and substantial economic uncertainty.[50]

48. F. L. Carmichael, "Housing in Denver: A Report on the Real Property and Low Income Housing Survey," *University of Denver Reports* 17 (June 1941), 25–26.

49. Denver Bureau of Public Welfare, *Denver Relief Study*, pp. 38–39.

50. Residents of Kansas City's Country Club District were likely to work as attorneys, physicians, and dentists or as managers of large firms in such industries as meat packing, railroading, and banking. For a list of buyers and their titles early in 1922, see

MATCHED CLUBS for HIM . . .

Ever ask HIM why he insists on a matched set of clubs when he plays golf with the boys on Saturday afternoon? He'll tell you it helps his game, gives him confidence, lowers his score. True enough, and if it's true on the fairway, it's doubly true in the kitchen.

The all-electric "matched kitchen" will help your budget, give you confidence with new recipes and save hours of time for things you like to do.

First, equip yourself with an electric refrigerator . . . to save food and provide plenty of ice for cool drinks this summer. Next, the electric range . . . now fast, automatic and economical. Then, get the electric water heater . . . for controlled hot water, instantly ready at the turn of the tap. And finally, the electric dishwasher . . . to eliminate the most irksome task of the kitchen.

GO MODERN, with an all-electric "matched kitchen" . . . for more skill in the fine art of living.

A MATCHED KITCHEN for YOU!

Fig. 14. "Matched Clubs for Him . . . A Matched Kitchen for You!" April 22, 1940. By 1940, gender assignment was part of virtually every sales promotion. (Source: *Power & Light News,* Records of the Kansas City Power & Light Company)

Equally, men occupying senior positions had begun to adopt a more relaxed stance toward their wives. One might speculate that a business community emphasizing associative activities and trust in the judgment of one's colleagues across the nation encouraged a large number of these men to begin to trust the judgments of their spouses. For whatever reason or reasons, a companionate marriage, one of friendly equals, appeared to be a desirable goal—or at least an attractive idea. The creation of an inner arena—the home—in which family members could communicate intimately with one another (perhaps with

CCD Bulletin 4 (June 1922), 2; and for essays on wealthy householders and their architects, who were designing homes that enhanced family togetherness and defensiveness against industrializing cities, see Robert C. Twombly, "Saving the Family: Middle Class Attraction to Wright's Prairie House, 1901–1909," *American Quarterly* 27 (March 1975), 57–72; and Sies, "City Transformed."

some consequent diminution of public responsibility) became important for these men. Yet they never doubted their obligation to protect wives and daughters from the dangers of industrializing cities. A home equipped with gas and electric appliances—up-to-date, free of smoke and drudgery, and located in an elite residential district—offered a congenial site in which companionship, familial devotion, and patriarchy could reside in a workable tension and in a physically secure setting. As early as 1923, then, Munroe could advertise gas furnaces, which were installed almost exclusively in the homes of the well-to-do, as heralding the day "when it is no longer necessary for either man or wife to be the janitor and stoker."[51]

Nichols and Munroe reached young adulthood during a period of remarkable growth and change in their cities. In the course of their work-lives, the trolley and then the automobile and truck assumed prominent roles in urban transportation, and gas and electric executives supervised the installation of countless miles of pipes and wires throughout their cities, offering unlimited service to all who could pay. In turn, the ubiquity of automobiles, trucks, and electric and gas utilities permitted affluent householders to relocate from smoky and congested areas near the central business district to distant neighborhoods.[52]

51. PSCC, *Invisible Furnaceman.* Historians have prepared a substantial body of literature focused on the origins and meaning of the companionate marriage. Karen Lystra, *Searching the Heart: Women, Men, and Romantic Love in Nineteenth-Century America* (New York: Oxford University Press, 1989), pp. 127, 234–237, finds the companionate marriage based on romantic relationships that were developing during the nineteenth century. On the other hand, Marsh, "Suburban Men and Masculine Domesticity, 1870–1915," p. 113, determines that romantic love had only "softened" patriarchy. After 1900, contends Marsh, men with secure jobs in large corporations and secure homes in new suburbs began to take a greater interest in child rearing and household decorations. For the development of these "'intensely domestic' families" in houses located in areas much like those constructed by community builders like J. C. Nichols, see Margaret Marsh, "From Separation to Togetherness: The Social Construction of Domestic Space in American Suburbs, 1840–1915," *Journal of American History* 76 (September 1989), 511. Equally, Mark C. Carnes, *Secret Ritual and Manhood in Victorian America* (New Haven: Yale University Press, 1989), p. 155, argues that values previously judged feminine, such as cooperation, "brought rewards in the workplace." (Marsh and Carnes strengthen their own essays through effective borrowing of concepts from one another; I first learned of Carnes's publications through the essay by Marsh.) For efforts to institutionalize cooperation among business leaders in the form of associationalism, see Ellis W. Hawley, *The Great War and the Search for a Modern Order: A History of the American People and Their Institutions, 1917–1933* (New York: St. Martin's Press, 1979), pp. 100–104.

52. Greg Hise, "Home Building and Industrial Decentralization in Los Angeles," *Journal of Urban History* 19 (February 1993), 95–98, summarizes the scholarship that

Such men as Roy Munroe and J. C. Nichols were among the countless agents of diffusion. They reinforced a set of messages about culture, gender, and technology that educators, architects, utility executives, politicians, home builders, and environmental reformers had been making since the late nineteenth century. As sales personnel rather than educators or politicians, however, Munroe and Nichols focused conversations on specific appliances and followed up among the unconverted with a hard sell. Such efforts encouraged wealthier householders to equate gas and electric appliances with comfort, convenience, environmental control, and the anticipated elimination of dirt and disease from their surroundings. Earlier efforts to educate Americans about proper lighting, healthful circulation of air, improved methods of cooking, or personal hygiene had as one of their points of termination the gas furnace, the gas hot-water heater, air conditioning, and shopping trips to spotless stores, such as Wolferman's on the Country Club Plaza. Wealthy men came to adopt gas and electric appliances as a method of protecting their families in relationships that were emphasizing companionship in a cozy and healthful suburban setting. Among the less affluent, the presence of gas and electric appliances allowed women to provide for the direct material needs of their families. The absence of sparkling appliances only alerted less wealthy and unemployed men once again to their own inability to protect their wives and children. As sociologist Charles A. Thrall reminds us, household appliances sold by such executives as Munroe and Nichols entered the household in a conservative fashion.[53]

In the process of selling gas and electric appliances that were clustering in neighborhoods of warm meals, well-ventilated rooms, and comfortable shops, Munroe and Nichols were setting in motion a series of unexpected developments at the gas and electric corporations. Up to 1920, a few managers supervised electrical operations, political liaison, and sales programs. By 1940, large staffs of experts at the gas and electric firms in both Denver and Kansas City directed operations in a number of specialized areas, and paid particular attention to social, political, and spatial changes in their cities. In brief,

---

places highways and utilities at the center of the deconcentration process, and another body of literature that highlights the importance of factory location in creating a decentralized urban region. In his study of spatial outcomes in Los Angeles, Hise determines that the previous location of manufacturers was paramount.

53. Charles A. Thrall, "The Conservative Use of Modern Household Technology," *Technology and Culture* 23 (April 1982), 175–194.

managers of gas and electric firms redesigned their organizations and technologies with a view to making them conform to fast-paced urban growth and stratified diffusion. Between 1920 and 1940, managers of gas and electric companies in Kansas City and Denver sought to urbanize operations.

CHAPTER SIX

Adapting to the City Again, 1920–1940

> Four hundred employees of the Kansas City Power and Light Company received diplomas last night for completion of courses they had taken in the last six months in the night school sponsored by the company. One fourth of the company's employees take advantage of the courses offered in the school. They range from such knotty mathematics as differential equations to public speaking and bookkeeping. Instructors are furnished by the company and the Lathrop Polytechnic Institute. The graduation exercises were held at Edison Hall in the Power and Light building. The commencement address was given by H. Roe Bartle, Kansas City Boy Scout executive. About 600 persons were present besides the graduating class.
>
> —*Kansas City Times,* May 19, 1936

DURING THE INTERWAR YEARS, salespersons like Roy G. Munroe and J. C. Nichols had achieved considerable success in selling appliances to far-flung householders. In turn, Munroe and Nichols were participants in larger processes of social change and technological diffusion. Among wealthier households, the interwar decades were the period during which suburban residence and ownership of new gas and electric appliances first emerged as prerequisites for a hygienic lifestyle, a companionate marriage, and middle-class standing. Gas and electric service were becoming embedded in social institutions.

Because social and urban change was so rapid in the years following World War I, executives of gas and electric firms believed that their technologies, employees, and organizations were lagging. Ordinarily, technical and financial experts employed at gas and electric firms broke problems into small parts and solved them. Yet problems associated with providing gas and electric service to residents of fast-changing cities looked "messy." During the interwar decades, then, executives of gas and electric firms in Kansas City and Denver sought to create a capability among ordinary employees for identifying social and urban change and for institutionalizing their perceptions.[1]

This process of adapting to urban and social change followed three overlapping courses. First, gas and electric managers educated their employees about the demographics and the political economy of the urban region and the company's role in it. Second, senior executives reorganized their firms, creating new divisions for specialized personnel. Third, top-level managers and engineers sought to anticipate (and perhaps to direct) the rapid outward movement of householders and businesses by constructing a high-voltage loop around the city far in advance of new subdivisions. Altogether, gas and electric managers took these three steps in order to align their organizations and technologies with changes in the social and economic geography of their cities and regions. During the interwar decades, executives of gas and electric firms in Kansas City and Denver completed the urbanization of their operations.

In virtually every North American city during these years, including Kansas City and Denver, populations increased in size and moved

1. For development of the concept of cultural lag on which my use of technological lag is based, see the Epilogue; and for a critique of cultural lag, see George H. Daniels, "The Big Questions in the History of American Technology," *Technology and Culture* 11 (January 1970), 1–21. For the metaphor of technological systems as composed of "messy . . . components," see Thomas P. Hughes, "The Evolution of Large Technological Systems," in Bijker, Hughes, and Pinch, eds., *The Social Construction of Technological Systems* (Cambridge: MIT Press, 1987), p. 51. Moreover, in his *Networks of Power,* p. 363, Hughes points to "a culture of regional power" being constructed during the 1920s around power grids that delivered high-voltage current over long distances. Altogether, Hughes speaks of "expanding power systems" and of "organizational forms" that, he finds, were capable of assembling and applying the specialized knowledge of science and technology. I have no quarrel with these conceptualizations. Hughes's publications have informed my research in concept and in detail, including, for example, his observation in *Networks of Power* that engineers "define solvable problems" (p. 14). My own emphasis, however, is on identifying the particular steps by which technologists perceived and defined the social and spatial composition of the urban region as factors that were independent, potentially disruptive, or even politically controversial. See also Susan J. Douglas, "Technology and Society" (review of Bijker, Hughes, and Pinch, *Social Construction of Technological Systems*) in *ISIS* 81 (March 1990), 80–84, for observations that were valuable in helping me define my own objectives.

Table 2 Populations of Denver and Kansas City (Mo.), 1920–1940

	1920	1930	1940
Denver	256,491	287,861	322,412
Denver metropolitan area	329,948	385,019	445,206
Kansas City, Mo.	324,410	399,746	399,178
Kansas City metropolitan area	475,141	613,954	634,093

SOURCE: U.S. Department of Commerce, Bureau of the Census, *Sixteenth Census of the United States: 1940,* vol. 1 (Washington, D.C.: Government Printing Office, 1942), pp. 32–33; Kansas City (Mo.) City Plan Commission, *The Kansas City Metropolitan Area* (1947), p. 45; Stephen J. Leonard and Thomas J. Noel, *Denver: Mining Camp to Metropolis* (Niwot: University Press of Colorado, 1991), p. 481.

away from downtown. Between 1920 and 1940, the population of Denver increased 25 percent, and that of Kansas City jumped 23 percent. Even more rapid was the rate of growth in the outlying areas. During the 1920s, reports Kenneth Jackson, "new suburbs sprouted on the edges of every major city." Naturally, areas adjacent to Kansas City and Denver participated in this rapid development. Between 1925 and 1928, Denver's far southside as well as fashionable Park Hill to the east of downtown were the fastest-growing sections of the city. Not even the harsh economics of the depression decade slowed the growth of distant suburbs. The boom years that had characterized the "instant cities" of Denver and Kansas City were several decades in the past. Yet neither an effort by political and business leaders in Denver to manage growth, or the leveling-off of industrial expansion in Kansas City, slowed population increases across an expanding region.[2] (See Table 2.)

Movement of businesses followed two paths. Large-scale manufacturers such as meatpackers who were dependent on rail transport for raw materials and nationwide shipment of finished products stayed put in industrial districts, mostly near downtown. Other businesses—including retailers who were following customers, and smaller manufacturers who were able to make use of trucks—began to scatter, creating lengthy commercial strips along their city's major thoroughfares. In Kansas City, commercial development on several streets

2. Kansas City, Missouri, City Plan Commission, *The Kansas City Metropolitan Area* (1947), p. 45; Kenneth T. Jackson, *Crabgrass Frontier: The Suburbanization of the United States* (New York: Oxford University Press, 1985), p. 175; "Building Operations in Denver Classified by Districts and by Type of Building," *University of Denver Business Review* 6 (January 1930), 6–7; Lyle W. Dorsett, *The Queen City: A History of Denver* (Boulder, Colo.: Pruett Publishing Company, 1977), p. 200; A. Theodore Brown and Lyle W. Dorsett, *K.C.: A History of Kansas City, Missouri* (Boulder, Colo.: Pruett Publishing Company, 1978), pp. 183–184.

extended as far south as Eighty-fifth Street. Not only did many employers in both Kansas City and Denver locate far from downtown, but they also took a number of jobs with them, adding to the flow of persons who traveled long distances on a daily basis between work and residence.[3]

Relocation of householders and businesses reinforced the arrangement of residential districts according to income, race, and ethnicity. In 1940, African and Mexican-Americans totaled 5 percent of Denver's population. But 76 percent of the city's black residents and 67 percent of the Mexican population resided in only two census tracts, both near downtown and in the old industrial district. Household income also was distributed unevenly across the urban landscape. In 1940, the average Denver household (a statistical fiction) earned $1,420 a year. More telling was the fact that residents of one census tract located near downtown earned $690 while those living in fashionable subdivisions located far to the east were enjoying an income of more than $2,900. And by the early 1940s, many of the wealthiest residents of Kansas City and Denver were moving beyond settled areas such as the Nichols Country Club District. Neither income disparities nor the clustering of householders according to race and ethnicity were new phenomena in American urban history. For that matter, gas and electric executives like Henry Doherty and Roy Munroe had long recognized that residents of different portions of the city consumed gas and electricity in uneven amounts. Between 1920 and 1940, rich and poor and black and white simply moved farther from one another.[4]

Rapid increases in demand for gas and electricity added another dimension to the complexities utility managers in booming cities faced. During the interwar decades, demand for electric and gas service throughout the United States increased nearly 320 percent.

3. Fred A. Lampe, "Land Use Patterns in Kansas City, Missouri: An Empirical Investigation" (Ph.D. dissertation, University of Kansas, 1972), pp. 72–75, 86–89; Kansas City, Mo., City Plan Commission, *Kansas City Metropolitan Area,* pp. 9, 11, 24–30; F. L. Carmichael, "Housing in Denver . . . " *University of Denver Reports* 17 (June 1941), 7, 30. For an account of a similar pattern of deconcentration in Philadelphia during this period, see Eugene P. Ericksen and William L. Yancey, "Work and Residence in Industrial Philadelphia," *Journal of Urban History* 5 (February 1979), 147–182.

4. Carmichael, "Housing in Denver," 15, 21. For the migratory patterns of Denver's highest-income householders and their children, compare "Denver Social Record and Club Annual" (n.p., c. 1927); *Social Record and Club Annual: Denver, Colorado, 1930–1931* (Denver: George V. Richards, 1931); *Social Record and Club Annual: Denver, Colorado, 1935–1936* (Denver: George V. Richards, 1936); and *Denver Social Record and Club Annual, 1941–1942* (Denver: Program Publishing Company, c. 1942).

Table 3 Use of Electric Energy in the United States, 1920–1940 (in millions of kilowatt-hours)

Year	Residential	Commercial	Industrial
1920	3,190	6,150	31,500
1925	6,020	9,345	45,500
1930	11,018	13,944	61,023
1935	13,978	13,588	63,265
1940	24,068	22,373	92,390

SOURCE: U.S. Department of Commerce, Bureau of the Census, *Historical Statistics of the United States, Colonial Times to 1970,* part 2 (Washington, D.C.: Government Printing Office, 1975), p. 828.

Table 4 Use of Gas Energy in the United States, 1920–1940 (in billions of cubic feet)

Year	Residential	Commercial	Industrial
1922	254 (combined)		509
1925	272 (combined)		916
1930	296	81	1,565
1935	313	100	1,497
1940	444	135	2,076

SOURCE: U.S. Department of Commerce, Bureau of the Census, *Historical Statistics of the United States, Colonial Times to 1970,* part 2 (Washington, D.C.: Government Printing Office, 1975), p. 831.

Changes in demand at the local and urban level were just as impressive. Between 1920 and 1939, the electric company in Kansas City increased production by more than 400 percent; and for the shorter period between 1923 and 1939, production at the electric company in Denver increased approximately 300 percent.[5] (See Tables, 3, 4, and 5.)

On the face of it, utility executives in Kansas City and Denver should have looked forward to serving a growing number of customers who were demanding increasing quantities of gas and electricity. But most gas and electric executives continued to operate under the same physical and financial constraints that had characterized their businesses from the start. Immense investments in distribution facilities to ser-

5. U.S. Department of Commerce, Bureau of the Census, *Historical Statistics of the United States, Colonial Times to 1970,* part 2 (Washington, D.C.: Government Printing Office, 1975), p. 828; supplements to *Electrical World,* April 24, 1926; May 3, 1930; May 9, 1936; May 4, 1940.

Table 5 Yearly Output of Electricity by Kansas City Power & Light Company and Public Service Company of Colorado, 1920–1939 (in millions of kilowatt-hours)

Year	KCP&L	PSCC
1920	175	162 (1923)
1925	339	244
1930	647	331
1935	652	354
1939	782	499

SOURCE: Supplements to *Electrical World*, April 24, 1926; May 3, 1930; May 9, 1936; May 4, 1940.

vice far-flung populations who demanded service on an occasional basis threatened financial disaster. Equally threatening was the prospect that consumers would demand light and heat mostly during cold and gloomy periods, leading to massive investments to service tall and slender periods of peak demand. As late as 1939, the amount of current demanded for "instantaneous" and thirty-minute peaks and the dates of those events were regularly reported within each firm and then distributed nationwide to industry leaders.[6]

During the early 1920s, executives at the gas and electric companies in Kansas City and Denver began to consider the implications for their own operations of household and industrial populations that were increasing in size and diversity, moving outward, and consuming greater amounts of gas and electricity. Executives of gas and electric companies in metropolitan areas had long taken account of political, industrial, and social change in their marketplaces. As early as 1882, operators of the first electric light companies in Denver and Kansas City had located their plants and distribution systems to secure access to coal and to provide service to nearby businesses. Similarly, during the period after 1900, Henry Doherty had adjusted rates and created a large sales force with a view toward social and political change in Denver. Mass production and volume sales were his responses to a fast-growing city and to politicians demanding area-wide service at low rates.

During the interwar decades, however, executives and senior members of their engineering and management staffs began to study their cities and regions in a truly systematic fashion. After 1920, the goal

6. Supplement to *Electrical World*, May 4, 1940.

of utility executives was to bring their operations into a closer and more predictable alignment with urban change. Education of personnel was requisite to creating the ability to shape and carry out policies that dealt in a routine fashion with rapid urban change and substantial increases in demand.

Up to the 1920s, education of employees of gas and electric companies for urban change had been uneven in focus and intensity. During the period between 1901 and 1917, Henry Doherty's executives in Denver and then nationwide—with their daily meetings, classes, required reading, and tests—had provided a consistent flow of technical, administrative, and urban information to employees. In contrast, leaders of gas and electric firms in Kansas City had never shaped a consistent policy regarding the education of employees. Beginning in 1887, Edwin Weeks, as the principal executive of the electric company in Kansas City, had encouraged several employees to offer courses in arithmetic, physics, and electricity. By 1897, however, rising expenses and rate wars encouraged Weeks to cancel the courses. In 1911, J. Ogden Armour reinstated a program of education, but canceled it two years later in favor of technical training at a local trade school. Nor had executives of the pipeline and distribution firms supplying gas to Kansas City appreciated the value of systematic education for urban change. Instead, they preferred to direct resources into what turned out to be exhausting battles over rates and service with customers, politicians, regulators, and plaintiffs.

By the early 1920s, only Doherty's large and successful organization had survived into the beginning of the decade. Not that formal courses of instruction had made the difference between success and failure for complex gas and electric firms operating in turbulent urban markets. Rather, education at Doherty's firms, including his gas and electric firm in Denver, was one of several important facets of operations aimed at encouraging closer coordination between huge investments, immense technical systems, and booming, diversifying, and decentralizing American cities. After 1920, leaders of new management groups at Kansas City's gas and electric firms implemented educational strategies for their own employees aimed at achieving a closer coordination with their cities.

Beginning in 1921, executives at Kansas City's reorganized electric company launched a large and sophisticated program of employee education. By the mid-1920s, educational activities at Kansas City Power & Light took place at three levels of social and political sophistication. At the first level, company employees studied matters that cannot be described as other than strictly technical. In Practical Electricity I, for example, students devoted approximately half their time to "transformer and rectifier principles" and the other half to arithmetic. In Practical Electricity II and III, the overall focus on technical matters remained in place, with students learning the details of Ohm's Law and how to calculate "the impedance of circuits containing inductance and capacity." In courses such as Practical Electricity, moreover, a situation euphemistically described by one company employee as "a wide variety of scholastic attainment" among students limited advanced study of topics such as Ohm's Law of circuitry. Particularly problematic, according to this observer, was the apparent fact that "students were greatly scattered out along the mathematics highway."[7]

Attendance at courses scheduled two nights a week and lasting two hours each night for a period of six months must have enhanced employee understanding of the equipment they handled during the day. Perhaps study by technical employees of transformers, circuits, meters, and mathematics would eventually foster improved understanding of electrical operations as a system. One must suspect, however, that focus on the technical details of circuitry by men assigned to tasks such as installing electric lines, repairing meters, and washing lamps would never afford them an opportunity to expand their comprehension of the emerging role of electrical operations in the city. Surely the course in sewing, taught and attended exclusively by women, could not foster an enhanced comprehension of the company's urban operations. For employees lacking the good fortune to enter employment with advanced educations (and especially advanced education in mathematics), these courses only reinforced the existing division of intellectual, political, and sexual labor.[8]

At a second level of social and political sophistication, however, employees of Kansas City Power & Light studied the interrelationship of their work with that of employees in adjoining units. The stated purpose of instruction for students enrolled in the course in power-plant engineering, as an example, was to learn "'how' and 'why' for the practical man." An engineer with the trolley company provided instruction in this course. An important component of the study program consisted of visits to "ten or twelve" power plants in the city. Only the most disinterested employee could have avoided recognizing that each of those plants carried a different load according to the peaks and valleys created by different groups of consumers located nearby.[9]

7. "First Annual Education Edition," *The Tie* 7 (March 1927), 11.
8. "Night School," *The Tie* 7 (December 1927), 13.
9. "First Annual Education Edition," pp. 11–12.

Still other courses emphasized the interrelationship of tasks in an explicit fashion. At the Foreman's Conference during 1925–1926, for example, members of the class brought forward "problems which arise in their work and the class analyzes the solution." According to a description of the course provided by a company employee, students discovered "a common interest." Similarly, employees attending the Workman's Compensation Act conferences for 1926–1927 acquired, according to one account, "a pretty fair working knowledge of the act as it applies to our company affairs." Again, the pedagogy of courses functioning at this inbetween level focused on aspects of electric operations that were often discrete in nature. Yet the net effect of lectures and discussions was to encourage students to perceive the interrelatedness of machines, situations, laws, and co-workers. One of the long-terms goals of education at Kansas City Power & Light, then, was "not only to give employees a broader knowledge of the company and its affairs, but also to introduce them to each other and also to recognize the talent we have here at home."[10]

At still a third level of sophistication, employees of Kansas City Power & Light learned of the relationships between their jobs and the welfare of the company in its social, economic, or political setting. On November 3, 1925, for example, participants in the educational program heard a lecture on limitations placed on the supply of direct current to the business district and the installation of automatic substations for the outlying areas. Movies were a popular component of this type of instruction, helping to develop the larger picture for employees. During the evening of November 9, 1926, workers viewed a double feature. The first, described as a comedy, dealt with life in the navy, and the second, prepared by General Electric, portrayed "the use of electricity to relieve the burden of farm labor." In December, students viewed a film prepared by Westinghouse dealing with railway electrification. Following the showing of these films, students, approximately 150 in number each evening, remained for refreshments and dancing. Inevitably, the goal of introducing employees to one another overlapped with instruction in the relationships between the company and its environment.[11]

Occasionally, instruction focused on the company and its external relationships was simply hortatory. During 1926–1927, instructors in the public speaking course "drill[ed] students" in the basic elements

10. Ibid., pp. 15, 17; and for an account of earlier development of these themes, see "Weekly Educational Meetings," *The Tie* 5 (November 1925), 1.

11. "Weekly Educational Meetings," p. 5; "First Annual Education Edition," p. 17.

of diction, aiming to prepare "public speakers ready to tell with eloquence the whole world the story of our utility and its product." One of the company's vice presidents explained his version of that story in a straightforward manner to employees. Formerly, he told them on October 26, 1926, the watchword of electric companies had been "The Public be damned," but in this era the key idea to keep in mind was "The Public Be Served." Within a few years, the idea of service had apparently filtered down to the level of ordinary employees. In June 1930, a member of the Right of Way Department delivered a lecture entitled "The Electric Utilities' Contribution to My Community, State and Nation."[12]

Whatever the pedagogic or social value of dances, films, or the simple retelling of "the story of our utility," classroom instruction often included hardheaded accounts of the company as part of an economic, social, and political complex located in a city and a region. Senior executives took care to deliver this aspect of instruction to their employees. Late in November 1926, for instance, the company's purchasing agent explained that operation of a testing lab fostered the ability of his employees "to work with the manufacturer to perfect a more suitable product." Early in January 1927, as a second example, one of the vice presidents outlined negotiations with the Phillips Petroleum Company that led to electrical service for oil producers located in the Texas Panhandle. During that same evening, another senior official led students through the details of power-plant economics. His lecture contained the well-known point that the size and number of boilers and turbines had to be related in a close fashion to average daily loads. Presumably, students were aware that load at the power plant was always a composite of household and business demands that varied during the course of the day, the seasons, and the state of the urban economy.

No lecturer presented the Kansas City Power & Light Company in its larger setting. One might even speculate that several among the senior staff understood only their own portion of that setting. Taken as a whole, however, the portrait sketched for employees seeking to understand the larger picture was that of a complex firm whose success rested on relationships extending back to suppliers, down to the turbine rooms in Kansas City, out to thousands of consumers, and south to booming oil fields.[13]

12. "First Annual Education Edition," pp. 12, 19; "Night School Students Completing Their Courses," *The Tie* 10 (June 1930), 12.

13. T. R. Harber, "The Control of Material," p. 23; Chester C. Smith, "Our Venture in the Panhandle," pp. 24–25; J. A. Keeth, "High Pressure and Turbine Equipment as

Education of employees at the gas and electric company in Denver followed the path Henry Doherty had fixed as early as 1901. By 1905, Doherty's executives regularly instructed promising employees to follow prescribed courses of study in such topics as sales, management, and accounting. Between 1906 and 1919, more than 600 employees, mostly engineers, attended one of Doherty's elite schools aimed at preparing junior managers for senior posts in the organization, including the Denver organization. In 1920, nearly 500 graduates of these advanced programs of study remained in Doherty's employ.[14] The presence of that group throughout the Doherty system created a sizable contingent of managers possessing years of training and experiences in common.

By the 1920s, programs to educate employees in the company's procedures had been in place for two decades. Yet Doherty's managers continued to insist that graduate engineers learn about the company through firsthand experience in departments. This period of training lasted about a year and a half. New employees with degrees in engineering began in the accounting department, where they took applications for new service, checked on credit references, and so forth. Thereafter, they moved to bookkeeping and then to such varied departments as meter-reading and collections. As an additional dimension of their studies and experiences, new employees were expected to prepare written reports of their activities. Supervisors graded those reports, with the idea that awarding grades would encourage students "to make careful observations . . . and to put these observations into written form in a clear, concise and correct manner." In turn, superiors expected a student to retain corrected reports for "later years, when he holds a position of responsibility and trust." Supervisors also graded students on "personality, conduct, initiative, earnestness, ability and adaptability, punctuality and other personal traits."[15]

Naturally, portions of work undertaken during this period of practical experience turned the attention of new employees toward the city and its surrounding region. As part of their training, every new employee spent four months in the new-business (sales) department.

Applied to the Present Northeast Station," pp. 26–29—all in "First Annual Education Edition," *The Tie* 7 (March 1927).

14. H. B. Shaw, "The Doherty Training Schools," *Doherty News* 5 (March 1920), 5–10, Munroe Files.

15. *Doherty Training Course Conducted by the Public Service Company of Colorado, Denver, for the Henry L. Doherty Company Organizations Operating Managers of Cities Service Company* (n.p., c. 1930), pp. 27–28, in PSCC Records.

Those assigned to Denver spent their four months under the supervision of Roy Munroe and his senior associates, learning at first hand the importance of stressing cleanliness, comfort, and convenience in sales presentations. During October 1930, for example, Munroe worked with five junior engineers enrolled in the training program. Alongside sales personnel enjoying years of experience in door-to-door sales, these recruits rode the trolley each day to meet likely prospects. During several days, they visited householders located in the fashionable Park Hill and Country Club districts. On other days, those junior engineers rode the trolley to meet residents of the packinghouse district northwest of downtown, a neighborhood inhabited mostly by recent arrivals from eastern Europe.[16]

This training program reinforced an evolving orientation toward the city's diversity. In 1919, for example, Arnold Perreten took part in the Junior Engineering program, in which he and his new colleagues "went from one department to another." Even during this period of training, he recalled, "we actually helped operate the department." During the next two decades, management assigned Perreten to positions as electric distribution engineer in Cheyenne, Wyoming; as superintendent of electric distribution in Boulder, Colorado; and in electrical distribution at the Denver property. In 1938, he moved to a supervisory position in distribution.[17]

The imperatives of managing a large electric company in a city stayed with Perreten throughout his career. Although he was involved exclusively with the company's technical systems, Perreten remained impressed with Denver's transportation system and political geography. "The streetcar system covered almost all of the Denver area," he remembered, and sales representatives as well as other employees used it "in their line of duty."[18] Altogether, it is not difficult to envision a young engineer suddenly achieving a heightened sense of awareness regarding the diversity of passengers on those trolley lines and the diversity of customers located within the Denver service area. At the conclusion of their periods of study and observation under the supervision of such executives as Roy Munroe, Doherty's junior employees (as well as their counterparts at the gas and electric companies in Kansas City) could not help but reinforce a growing appreciation among senior staff of the idea of restructuring their firms to take account of urban change. The idea underlying these reorganization

16. Ibid., p. 32.
17. "Interview with Arnold Perreten," c. 1975, PSCC Records.
18. Ibid.

plans was to institutionalize the urban orientation of employees at every level, including their orientation to the habits of men and women and the social geography of race and income.

During the 1920s, executives of large firms such as Sears, Roebuck & Company and the General Motors Corporation continued to restructure their organizations to enhance the "visible hand" of management. In particular, they sought to control vast quantities of raw materials and finished products traveling rapidly between fields, mines, factories, distributors, and consumers. In addition, executives of General Motors and Sears, among other firms, relied increasingly on the services of well-trained specialists to handle each of these operations in a predictable fashion, further encouraging the creation of decentralized firms built around line and staff systems. In turn, innovations in the structuring of these large firms rested on years of experience and experimentation in the organization of management and labor extending back to the railroad industry in the middle of the nineteenth century.[19]

After 1920, the exciting idea of restructuring firms to enhance managerial efficacy made its way into the deliberations of leaders of the nation's gas and electric firms. Similar to counterparts in highly capitalized and integrated companies such as General Motors, specialists at large gas and electric firms directed operations in such diverse fields as accounting, finance, marketing, pipeline construction, electrical manufacturing and distribution, and appliance and machinery sales and service. In addition, gas and electric officials trained staff members who dealt in a routine fashion with several hundred thousand customers possessing vastly different experiences as members of the city's many ethnic, racial, and income groups.[20]

In 1925, senior officials of Henry Doherty's vast Cities Service Company took control of the Kansas City Gas Company. Until the mid-1920s, managers and ordinary employees of the old company had carried with them the accumulating ill-will of more than two decades of service judged inadequate and of prices perceived as too high. Under Doherty's supervision, goals of day-to-day operations now included "improve[d] . . . service to our customers . . . [and] doing

19. Alfred D. Chandler Jr., *The Visible Hand: The Managerial Revolution in American Business* (Cambridge: Belknap Press of Harvard University Press, 1977), pp. 455–476.

20. Compare with Chandler's observation in ibid., p. 204, that electric, gas, and trolley systems "remained smaller and less complex than the older railroad systems."

everything we can to be good citizens in the community which we are privileged to serve."[21]

Doherty's executives in Kansas City organized the firm into three general departments—operating, accounting, and new business, with the latter unit responsible for the sale of domestic appliances and industrial machinery. The operating department, headed by a chief engineer and a number of supervising engineers, handled the distribution of gas, appliance installation, and appliance service. In addition, executives of the operating department created a school to train crews in appliance installation and service. Creation of a school to instruct crew members in the workings of a large number of appliances also permitted managers to transfer personnel according to seasonal demands and the short bursts of increased work that accompanied promotional efforts.[22]

Because Doherty and his top officials in New York insisted on the centrality of appliance sales to the sale of gas (and electricity), the new business department assumed a central role in Kansas City's reconstituted gas company. Doherty assigned several of his senior officials from other properties to hire and train representatives and to direct massive promotional efforts. One of those executives, H. C. Porter, had started his career in 1906 at the Denver Gas & Electric Company. As late as 1941, he recalled with pride and approval his first report to Doherty on the importance of load-building, knowledge of company operations, and "public relations."[23]

Doherty appointed F. M. Rosenkrans to the post of new-business manager, the position Roy Munroe held in Denver. Rosenkrans was a graduate of the University of Wisconsin with a degree in electrical engineering and had also launched his career with the Doherty organization in Denver. During his early years as an engineer with Doherty's growing company, managers transferred Rosenkrans to properties in Toledo, Ohio; Pueblo, Colorado; Tonawanda, New York; and throughout their holdings in Canadian Natural Gas. Following service during World War I in the U.S. Army Corps of Engi-

21. T. J. Strickler, "A Record of Progress—The Kansas City Gas Company," *Gas Age* 88 (July 31, 1941), 52–53.

22. Ibid., p. 52; C. H. Waring, "Our Distribution System," *Gas Age* 88 (July 31, 1941), 56–58.

23. H. C. Porter, "Yesterday and Today in Selling," *Gas Age* 88 (July 31, 1941), 60. On the centrality of appliance sales among directors of the Doherty organization, see Thomas F. Kennedy, "Important Factors Which Point Out 'the Trend of the Load,'" *Gas Service* 1 (March 15, 1926), 2–3. I am pleased to acknowledge access to early issues of *Gas Service* at the office of the Gas Service Company in Kansas City, Missouri.

neers, Rosenkrans was assigned to a senior post at corporate headquarters in New York City. With the consolidation early in the 1920s of Doherty's gas operations in the mid-continent region, Rosenkrans assumed control of new business in Kansas, Missouri, and Oklahoma. Starting in 1925 and continuing after World War II, Rosenkrans was the executive responsible for creating an organization capable of translating Doherty's traditional goals of employee education, public relations, and load-building into solid results for the new property in Kansas City.[24]

As early as 1926, an executive reporting to Rosenkrans told gas industry executives at a national meeting that the key to boosting the sale of gas for industrial purposes was to ascertain "the character of the city." In part, recognition of the city's character amounted to collecting data that highlighted the economic composition of Kansas City, particularly in terms of the potential sale of gas to different types of companies. According to this reasoning, gas executives were to conduct surveys of hotels, restaurants, bakeries, and other companies that used gas in large volume for heating and manufacturing. By 1941, following more than a decade of industrial sales work, another of Rosenkrans's subordinates reported once again the importance of recognizing the industrial composition of each city. Naturally, new categories of gas consumers—including manufacturers of beer, breakfast foods, and ice cream—had joined the list alongside restaurants and hotels. In Kansas City and elsewhere, moreover, sales personnel specialized in certain types of customers. During the interwar period, then, creation of an industrial classification scheme as well as a sales force that matched that scheme began to emerge as routine practice in the Doherty organization.[25]

Attention to Kansas City's large immigrant population was another part of the growing recognition of the city's "character" and the need to shape an organization to conform to it. Although Doherty along with other successful executives, such as Samuel Insull, had long insisted on door-to-door sales efforts, they had not created specialized appeals for each segment of the population. Those not able to read English were often overlooked in sales programs. Beginning around

24. "Has Been Through the Mill," *Gas Service* 1 (October 25, 1925), 5; F. M. Rosenkrans, "Sustained New Business Program Brings Results," *Gas Age* 88 (July 31, 1941), 27–30.

25. D. A. Campbell, "The Organization and Operation of an Industrial Gas Sales Department," in American Gas Association, *Proceedings of the Eighth Annual Convention* (New York, 1926), pp. 762–763; J. I. Carmany, "Industrial Gas Becomes Increasingly Important," *Gas Age* 88 (July 31, 1941), 35–36.

1905, however, Doherty and his sales executives, such as Roy Munroe, had begun to crystallize the idea of the city as a series of residential and commercial districts. Rosenkrans had learned the identical lesson in Denver and had it reinforced during the course of his assignments throughout North America. By August 1929, the advertising manager for the Kansas City Gas Company launched another gas house-heating promotion. In anticipation of this promotion, he placed advertisements in newspapers catering to the city's black, elite, and laboring populations, in other newspapers published in Italian and German, and in still other newspapers that he described as catering to Jewish, Roman Catholic, and Protestant residents of Kansas City.[26]

Rosenkrans also presided over the reorganization and expansion of a home service department that was designed to give greater place to routine presentations to female customers. Beginning as early as 1901 in Doherty's organization (and around the same time in most large gas and electric sales organizations), men sold appliances to women. Women assisted the men by serving as home service agents. They instructed women customers in using appliances through cooking demonstrations held for groups in churches and halls or in the home. Between 1901 and the 1920s, moreover, executives of gas and electric companies throughout the nation shaped this division of labor into specialized departments employing women who had taken college-level courses in home economics.

During the late 1930s, Rosenkrans imposed still another dimension on the organization of his home service unit. In July 1937, according to a report published in a national trade magazine, two of Rosenkrans's senior executives "decided to undertake a new plan of customer relations through the home service department." They hired five additional women, bringing the total of female staffers to eight. Consistent with established practice in this area of sales, the director and most members of her department held degrees in home economics or "had home service education in a university." Publicists for *Gas Age* described the efforts of these women as "customer good will

26. Ray T. Ratliff, "Kansas City Launches a Gas House-Heating Campaign," *Gas Age Record* 64 (August 3, 1929), 163–164; Ruth Schwartz Cowan, "Immigrant Women and Electric Households" (paper presented at the Annual Meeting of the Society for the History of Technology meeting jointly with the Thomas Edison Symposium, Newark, New Jersey, October 1979). This paper first alerted me to the preference of sales managers at the nation's electric and gas companies for dealing with native-born persons. I want to thank Professor Cowan for sharing this paper with me.

builders [who] also try to interest the home maker in new methods of modern cookery."[27]

An important dimension of the home service department was the emphasis it placed on building good will among women customers. Managers aimed cooking demonstrations exclusively at female groups, such as Girl Scouts and religious organizations. Home service agents often prepared lectures on such exotic topics as China, including exhibitions of Chinese shoes, pictures, jewelry, and wedding gowns. At the conclusion of a lecture, agents followed up with a practical demonstration on methods of preparing Chinese cookies and chop suey on a gas range while wearing Chinese apparel. Location of the air-conditioned "Hostess Kitchen" in one of the stores in J. C. Nichols's Country Club Plaza expressed in symbolic, literal, and spatial terms the gender and class dimensions of home service work.[28]

Race also constituted a prominent feature guiding the activities of the reorganized home service department. During the 1930s, public and private agencies in Kansas City maintained separate facilities for black and white citizens and customers. Work of the home service department was organized in an identical fashion. Talks at the lovely and comfortable Hostess Kitchen were clearly aimed at white women, especially those possessing above-average incomes. Beginning in mid-1937, the director of the home service department offered lectures on gas cooking at the segregated Y.W.C.A., talks aimed exclusively at black women who worked as maids. By early 1938, approximately forty persons were attending lectures in each of two classes that met for an hour and a half a week over twelve weeks.[29]

Adaptation to fast-growing cities also extended into the technical sphere. Following World War I, utility executives nationwide sought methods to match their expensive and complex distribution systems with urban growth and diversity. Whereas urban scholars were beginning to employ ecological metaphors, such as "residential clustering" or "population succession," utility executives conceived of their operating regions as composed simply of geographic and political subdivisions. They spoke confidently about the "central business district," the nearby "industrial district," and the fast-growing and often affluent

27. "Kansas City Finds Home Calls Aid Customer Relations," *Gas Age* 81 (March 31, 1938), 26.

28. Dorothy Clock-Diggle, "Follow Each New Range Purchase with a Home Service Call," p. 42; and Fred C. Mueller, "Denver Women Learn Gas Story from Hospitality House," pp. 39–40—both in *Gas Age Record & Natural Gas* 79 (April 29, 1937).

29. "Kansas City Finds Home Calls Aid Customer Relations," 27.

"suburbs." By the mid-1920s, for example, an executive with the Northern States Power Company in Minneapolis outlined for other executives the "characteristics of a suburban load." Homes in one suburb near Minneapolis, he reported, were larger, more expensive, and farther apart than others in his region. Homes in that district were also "highly saturated with electric appliances."[30]

During the early 1920s, new managers at the electric company in Kansas City began to adapt these concepts to the organization of their distribution systems. In 1922, two engineers reported discovery of out-migration and settlement patterns in the Kansas City region. Rather than describing Kansas City as a collection of districts, however, they described it as a group of "centers." As a guide to shaping the company's electrical distribution system, those engineers recommended that officials conceive of supplying electric current to "these centers and their surrounding zones . . . [as] exactly the same as supplying a group of small cities spread through a large countryside." The idea, then, was "to treat the power supply in exactly the same way as if these centers were separated by considerable distances."[31]

Downtown was clearly one of those centers of electric consumption. By the mid-1920s, the president of Kansas City Power & Light had determined that continuation of the policy of "promiscuous extensions of the Edison System" of direct current serving the area in and around downtown would require "a tremendous investment in substation equipment." Additions to that system, he decided, "would be limited to the congested portion of the business district where the load density was high." Because direct current was (and remains) the preferred form of power for high-torque, variable-speed motors, managers of buildings with such equipment as elevators would presumably locate near direct current lines.[32]

30. R. R. Herrmann, "Characteristics of a Suburban Load," *Electrical World* 88 (December 18, 1926), 1259.

31. A. E. Bettis and C. W. Place, "Automatic Control of A. C. Distribution," *Electrical World* 80 (November 25, 1922), 1151–1153; "Functioning of a Power Sales Department," *Electrical World* 89 (April 22, 1927), 703–704. See also H. H. Kuhn, "Economic Generation of Electricity by Power Zones," *Electrical Review* 78 (April 30, 1921), 683–687, for a Kansas City Power & Light executives' early conceptualization of electrical distribution based on long-distance loops in large geographic regions. His reasoning, common in the industry, was that the peak load in towns located throughout the region would not occur at the same time as the peak in Kansas City. Diversity was the key to achieving a desirable load factor.

32. Joseph F. Porter, "The History of the Electrical Industry in Kansas City and Vicinity" (thesis, Iowa State College, 1934), p. 61; Bettis and Place, "Automatic Control of A.C. Distribution," 1151–1153.

Equally important from the point of view of providing electricity to residents of a city portrayed as a series of centers was the installation of a long-distance service loop. These loops, which provided alternating current to residential and industrial accounts, were quickly emerging as a favored method among utility executives in a number of cities for providing electric service to far-flung areas without enduring the immense costs of constructing additional generating plants. By 1926, engineers at Kansas City Power & Light had constructed a ninety-two-mile service loop running both north and south of the city from downtown. The loop constructed around Kansas City, moreover, extended far beyond existing population clusters; company officials hoped to anticipate growth. Creation of residential and business clusters, the thinking went, might even occur in an orderly and predictable fashion.[33]

Before 1920, only Doherty's executives in Denver had begun the tasks of educating personnel to social and urban change and creating an organization capable of adapting to their recommendations. By way of contrast, gas and electric operators in Kansas City had failed in most respects to institutionalize methods for guiding their organizations during periods of upheaval. In 1900, J. Ogden Armour had purchased the city's electric company from the original owners. During the next few years, he spent a large sum developing the company's generation and distribution facilities. Yet Armour never devoted much attention to cultivating the city's business and political leadership with lower rates and vast promotional campaigns. Nor had he built an organization staffed by persons capable of identifying rapid urban change and translating it into technical, policy, and organizational innovations. In the short run, the political and business failures of Armour's trolley firm had drawn his electric company into bankruptcy. The real failure in the electric business had been Armour's inability to integrate electric operations with the city's evolving social and political geography.

33. *Kansas City Star,* January 2, 1927, clipping files, Kansas City Public Library; Raymond C. Miller, *Kilowatts at Work: A History of the Detroit Edison Company* (Detroit: Wayne State University Press, 1957), pp. 278–284, describes a high-voltage loop constructed in the Detroit area. In 1910, Samuel Insull installed a service loop in the Chicago area as part of a strategy to improve the diversity factor of his operation. Thereafter, construction of loops became a common practice, especially in stations owned by his Middle West Utilities Company. See Forrest McDonald, *Insull* (Chicago: University of Chicago Press, 1962), pp. 137–143. See also Thomas P. Hughes, "The Electrification of America: The System Builders," *Technology and Culture* 20 (January

Managers of Kansas City's pipeline and distribution companies had also failed to adapt to social and political change. In 1906, with the natural gas pipeline nearing the city, operators of the local distribution company accepted a franchise that saddled them with low rates over a ten-year period. Worse yet, the pipeline operator could not deliver adequate supplies after the first couple of years of booming and often wasteful sales. During World War I, moreover, the gas company and the electric company lacked facilities to deliver adequate supplies for a city and a nation at war and for an urban population increasingly demanding light and heat from central suppliers at modest prices. Up to 1920, then, operators of the electric and gas companies in Kansas City had suffered from the histories of their earlier managers, particularly the inability of those managers to create organizations capable of identifying and interpreting the urban and social contexts of technical operations.

After 1920, new groups of managers at Kansas City's gas and electric firms launched self-conscious programs to adapt their organizations, employee training, and distribution systems to social and geographic change. At the electric company, the key conceptual breakthrough came in 1922, as two engineers identified the city as a collection of "centers." Yet it was also the case that senior managers could not rely on a few perspicacious engineers to guide day-to-day operations in growing and changing cities. Instead, managers launched large-scale programs of employee education and training. One goal of those programs was to upgrade the technical skills of employees in courses such as Practical Electricity. The essence of education programs, however, was to suggest to employees the degree to which they had to remain alert to urban and social change.

These programs of employee education rested on two assumptions. The first was that eventually students would perceive themselves and their companies as actors in urban politics, as participants in the process of deconcentration, and as agents of diffusion in identifying appliances with the gender and racial conventions of their city and nation. A second assumption managers made was that their organizations were capable of translating the insights of ordinary employees, such as F. M. Rosenkrans, into concrete programs.

Adaptation of gas and electric operations in Kansas City to urban and social change was part of a nationwide pattern of reorganization and change among utility operators. Increasingly, gas and electric

---

1979), 149–151, which initially called the importance of the service loop to my attention.

operators in that city behaved like their counterparts in Denver and elsewhere. On the surface, utility operators in both cities were creating uniform methods of operation, bringing managerial expertise to bear on great volumes of gas, electricity, and appliances moving rapidly from plants to thousands of consumers in houses, apartments, offices, and factories. Certainly, national meetings and industrial publications facilitated the emergence of this common style of organization. At their conferences and in their trade literature, experts outlined recent innovations in areas such as rate-making, industrial heating, or power sales. Periodicals such as *Electrical World* published more than 1,000 pages yearly, presenting in detail the experiences of utility managers with service loops, educational programs, organizational formats, and sales campaigns. The fact that each of these local companies was only one of many properties in the holdings of nationwide corporations like Doherty's gave managers access to the accumulated experiences of leaders occupying identical slots throughout North America.

Personal ties also facilitated the rapid diffusion of uniform methods. In the Doherty organization and later at Kansas City Power & Light, future executives met during training periods and strengthened those ties as they shared common responsibilities in new assignments. During the interwar decades, for example, Rosenkrans and Munroe exchanged sales data and practical information on certain topics including the cost, durability, and consumption levels of gas-fired furnaces and hot-water heaters. Munroe and Rosenkrans were at once allies and competitors in the business of creating and adjusting complex organizations.

Managers also carried most of the responsibility for the demanding task of adjusting those organizations to local and urban environments. Early on, Doherty and Munroe had recognized, virtually on an intuitive basis, that the city's politics, public policies, and social geography fixed the parameters within which managers made choices regarding gas and electric technologies. During the 1920s and 1930s, the task of such managers was to shape organizations that could routinely perceive that urban place and translate its many convictions regarding race, gender, and the desirability of moving to the periphery into shrewd corporate policy. During the interwar period, Doherty further systematized that effort at the combined gas and electric company in Denver and at the gas company in Kansas City. New managers at Kansas City Power & Light quickly adopted similar methods and organizational formats.

By about 1940, gas and electric operators in cities such as Kansas City and Denver sold gas and electricity in vast quantities to hundreds of thousands of businesses and householders. The comparative success of these organizations in completing the adaptation process directs attention to a related dimension of organizational change. Neither the management groups at the nation's urban gas and electric companies, nor the politicians, householders, or small and medium-size industrial firms in most cities had incorporated into their thinking carefully constructed models of urban diversity and organizational form. Instead, managers created metaphors and symbols that oriented them to the city and that facilitated their efforts to make sense of urban change. One metaphor was "center," which allowed managers and engineers at the electric company in Kansas City to conceptualize a long-distance transmission system. As a popular metaphor for most utility executives, however, the idea of a city as a series of centers lacked appeal.

Service was a much more useful notion. As early as 1910, Doherty's selection of Cities Service as the name for his nationwide company symbolized the careful attention to the urban context of operations expected of his managers. During the 1920s and 1930s, moreover, a rhetoric of service emerged as a routine feature in the language of utility managers. In part, recitation by executives of phrases such as "The public be served" reflected a concern for public relations. By the late 1930s, however, utility managers regularly reported to one another on their efforts to develop power service, customer service, home service, or automatic hot-water service. Service was an idiomatic expression among industry executives that allowed them to orient their firms toward the city, including its morphology, its economic diversity, its politicians, and the evolving perceptions of managers, employees, and urban residents about race and gender. By about 1940, the urbanization of gas and electric operations was essentially complete.[34]

Beginning after World War II, gas and electric operators throughout the nation built on their earlier experiences in developing urban markets. They spoke of cleanliness, comfort, and convenience, espe-

34. Anson D. Marston, "Why I Buy My Power Service from a Power Company," *Power & Light News* 6 (March 17, 1939), supplement; Gordon B. Koch, "Commercial Cooking Load Held by Customer Service," *Electrical World* 112 (September 9, 1939), 50; Clock-Diggle, "Follow Each New Range with a Home Service Call," p. 41; American Gas Association, *Proceedings of the Twenty-Second Annual Convention* (New York, 1926), p. 763. For comparable organizational changes at the American Telephone & Telegraph Company leading to a strategy of technical innovation and social and political adaptation over the long run, see Louis Galambos, "Theodore N. Vail and the Role of Innovation in the Modern Bell System," *Business History Review* 66 (Spring 1992), 95–126.

cially for women. For about twenty-five years following the war, utility operators throughout the nation converted perceptions of cleanliness and comfort for women (and men) into fabulous increases in the production and consumption of gas and electricity. During this remarkable period in the national experience, the imperatives of volume production of gas and electricity cohered with rising incomes, improved distribution systems, and an acquired taste for lots of light and heat. Among a large number of Americans, discussions of comfort and convenience at home assumed a transcendent quality, as if politics and public policy no longer mattered. Gas and electric executives, it appeared, had found a way to avoid the snares of politics and public policy. No wonder that during the 1950s and early 1960s, publicists and scholars alike could speak of an end of ideology.[35]

35. Daniel Bell, *An End of Ideology: On the Exhaustion of Political Ideas in the Fifties* (Glencoe, Ill.: The Free Press, 1960).

CHAPTER SEVEN

Baptismal Tanks and the Feminized Search for Environmental Perfection, 1945–1985

> By June 1944, appliances topped the list of the most desired consumer items. When asked what they hoped to purchase in the postwar years, Americans listed washing machines first, then electric irons, refrigerators, stoves, toasters, radios, vacuum cleaners, electric fans, and hot water heaters. Advertisers claimed that these items constituted the American way of life that the soldiers were fighting for.
>
> —Historian Elaine Tyler May, 1988

IN JULY 1950, A WRITER for *Architectural Forum* described a new church located in Minneapolis. The writer highlighted the heating and ventilating systems as well as "contemporary lighting methods . . . used to create the climax of the whole interior." So impressive was the mix of aesthetics and "environmental controls," contended the writer, that "art, science, and faith achieve a serene harmony in this simple church."[1]

During the 1950s, a search among church members and church architects for their own mix of science, faith, and environmental control achieved routine form in specifications for new buildings. Church architects prescribed installation of gas-fired furnaces and electrically operated air conditioners capable of heating main sanctuaries and

1. "Christ Church," *Architectural Forum: The Magazine of Building* 93 (July 1950), 80.

adjoining rooms to 70 degrees during the chilliest days of winter and cooling them to 75 degrees during summer's hottest days. The architect for one church in Kansas City instructed contractors to install a drinking fountain that would cool water to 50 degrees. By 1959, the search for a comfortable water temperature at the Red Bridge Christian Church in Kansas City also extended to their newest members. The architect specified installation of a gas-fired hot-water unit that was capable of bringing water to a temperature between 80 and 85 degrees, exclusively for the baptismal tank.[2]

That baptismal tank represented more than a symbol of faith or the product of art and science. The presence of gas-fired baptismal tanks, along with such innovations as chilled water and air conditioning, represented the increasing ability of ordinary Americans of every faith to regulate portions of their environments beginning soon after birth and ending only with death. Those heated baptismal tanks also highlighted the success after World War II of political leaders, engineers, and agents of diffusion in delivering inexpensive gas and electricity throughout the United States. Beginning around 1970, however, changes in the nation's political economy and in the technical ability of engineers to supply cheap and abundant energy led to dramatic increases in the cost of light and heat. Only the agents of diffusion persisted. Despite these price increases, the emphasis among home economists, educators, and others remained on products that were designed to enhance the comfort and convenience of women. (Because the postwar years were characterized by regional and national patterns of energy production and consumption, the focus of this chapter shifts from Kansas City and Denver to developments throughout the United States.)

After World War II, the process of informing ordinary Americans about gas and electric appliances got under way once again. During the last half of 1945, the editors of *Architectural Record* provided a forum in which advertisers and architects began to prepare readers for the postwar world of environmental control. Although a few authors and advertisers wrote in a nearly utopian language, with titles

2. "Specification of a Building for 39th and Flora Church of Christ," 1955; "General Specifications for Benton Baptist Combination Church and Educational Building," August 12, 1955; "Avondale Methodist Church," 1958; "Building for Red Bridge Christian Church, 109th and McGee Street," 1959; "Specifications for Mechanical and Electrical Work, Buildings for Saint Joseph's Parish, 17th and the Paseo, Kansas City, Missouri," 1960; "All Saints Episcopal Church," c. 1961—all at the University of Missouri at Kansas City.

such as "Highway Restaurant for a 100-Octane World," most were content to provide highlights of existing systems similar to those that would soon deliver environmental control to everyone on the spot. For example, designers of Thompson's restaurant in Chicago, described as catering to "the hungry worker and the wee-small-hours wanderer," included air conditioning along with fluorescent lighting "chosen in an assortment of colors to give the most natural possible color to the various foods."[3]

Not only could workers in local restaurants now enjoy their meals under natural lighting and at a comfortable temperature, but soon householders, especially women, who would be able to prepare interesting meals at home, would also benefit. Executives at General Electric advertised that "the appliances most women want most" included "automatic heating with air conditioning" as well as an "all-electric kitchen." Manufacturers of gas equipment advanced the same argument. "As soon as materials are available," promised the association of gas appliance and equipment manufacturers, "women who have already selected gas as best for cooking will have" ovens featuring pilot lights, "smokeless-broilers—simmer-speed top burners—[and] high-low temperature [for] exactly controlled ovens."[4]

Suppliers of lighting, heating, and ventilating equipment were equally alert to a potential market in schools, offices, and factories. In July 1945, Janitrol, a maker of gas-fired heating equipment, emphasized "heating comfort plus classroom beauty" along with "constant temperature for the storeroom." A manufacturer of oil-burning equipment, citing the authority of an architect working in New York City, contended that their unit "provides uniform heat that promotes both comfort and health." On the grounds that "your new schools will be no better than their ventilation," the John J. Nesbitt Company argued that their unit "mixes room air as desired and circulates approved quantities of clean, fresh, comfortable air." Whether in the buildings of tomorrow or in those planned for construction as soon as the economy returned to normal, the emphasis, as the manufac-

3. "Highway Restaurant for a 100-Octane World," *Architectural Record* 98 (October 1945), 102. See also, for a futuristic tone, "Catholic Church for the Orient," ibid. (September 1945), 99; and "New Notes Brighten an Old Name," ibid. (October 1945), 117.

4. "And Inside It Will Have . . ." (advertisement), *Architectural Record* 98 (October 1945), 62; "She Knows What She Wants and She'll Have It" (advertisement), ibid. (September 1945), 54.

"That's what I've been waiting for"

. . . and aren't you glad you waited! For when that big, handsome new Maytag comes through your door, wash-day drudgery flies out the window. Biggest washings, heaviest fabrics, grimy work-clothes, filmy lingerie, baby things . . . all are handled in a jiffy . . . in the Maytag's big, square cast-aluminum tub. And Maytag's exclusive gyrafoam action is so rough on dirt, so gentle on clothes! See this great new Maytag with its many new-day features at your dealer's now. The Maytag Company, Newton, Iowa. Washers . . . Ironers
Home Freezers Dutch Oven Gas Ranges

MILLIONS OF WOMEN HAVE THEIR HEARTS SET ON A NEW Maytag

Fig. 15. "That's what I've been waiting for," c. 1946. (Courtesy, Maytag Company)

Company Connected 200,000th Electric Customer on September 1

PUBLIC SERVICE COMPANY of Colorado marked an important milestone in the growth and progress of the territory it serves on September 1, 1950. The 200,000th electric customer was added to its system—the family of Mr. and Mrs. Lawrence Desmond, 3016 Cook street, Denver.

President J. E. Loiseau visited the Desmond's newly-built home to congratulate the couple who, unknowing, had attained the unique distinction when they signed up for service for their house, late in August. At the same time, Mr. Loiseau surprised and delighted the Desmonds with the gift of a new automatic electric washer to add to their home equipment and furnishings. Mr. and Mrs. Desmond are parents of two children, Tommy, 9 months and Mary Theresa, 25 months. Mr. Desmond, a driver-salesman for the Meadow Gold Creamery Company, is an ex-Navy man, having seen service in the South Pacific during World War II.

The company, up until September 1, 1950, had added 58,439 new electric customers since 1940.

The growth in the number of gas customers served by the company has almost kept pace with the growth of electric customers. The first gas customer to be served in Denver was connected in 1870, and at the end of 1940 the number was 85,959 customers. The 100,000th gas customer was connected during the year 1945. By the end of 1949 the number had reached 132,270.

One and all, the Desmonds beam with delight when advised of the company's gift of an automatic electric washer for their new home.

How to Become a Beauty Queen

GIRLS, here's a tip for free. They pick beauty queens these days by electric applause meters. If you want to win a beauty contest, pick relatives, friends and beaus with loud voices and strong hands.

BELATED STORK NEWS

Mr. and Mrs. K. L. Swynenberg, Denver Gas Engineering, a daughter, Sherry Lynn, on June 20th.

Mr. and Mrs. W. J. Dorland, Denver Gas Engineering, a son, Ralph Eugene, on July 30th.

Page 4 August-September, 1950

Fig. 16. "Company Connected 200,000th Electric Customer on September 1," August–September 1950. Following World War II, electric (and natural gas) service assumed routine form in the lighting, heating, cooling, cooking, and other household activities of many Americans. (Courtesy, Public Service Company of Colorado)

turer of a baseboard radiation unit put it, was on "No cold corners. No hot spots. No hot or cold levels."[5]

During the first two years following World War II, however, large numbers of Americans lacked access to wonderful machines delivering comfortable lighting and temperatures. In part, the inability of manufacturers to supply domestic appliances following production for war forced Americans to wait. In 1946, Sears, Roebuck & Company advertised home freezers and vacuum cleaners along with a number of other items in their catalogs. This merchandise was not yet in stock, but executives remained confident of shipments within a few months. Those shipments never arrived, however, and the company had to make refunds in the amount of $250 million. During 1947, Sears executives did not even bother to advertise hard-to-find products such as stoves, sewing machines, and electric refrigerators. As late as March 1947, then, the author of an article instructing consumers on the uses of heating equipment reported: "The lucky purchaser of a reliable pre-war type of boiler, burner, or stoker will not be likely to find his new equipment outmoded for several years to come, at least."[6]

The postwar housing shortage added to the inclination of many Americans to postpone appliance purchases. "Shortages of materials and labor and the high costs of both," ran the reasoning among editors of *Popular Mechanics* in April 1947, had created "the most serious housing crisis ever experienced in America." With mortgage insurance provided by the federal government through the Federal Housing Administration and the Veterans Administration, housing starts zoomed upward, reaching a record 1,692,000 in 1950. Even this rapid pace of construction was not sufficient for the vast numbers of veterans and others seeking a single family home. Indeed, reports Kenneth Jackson, a trend begun during the Depression of moving in with family or friends extended into the postwar period, with 6 million

5. "6 Common Heating Problems Solved at Spartan by Janitrol" (advertisement), *Architectural Record* 98 (July 1945), 13; "Your New Schools Will Be No Better Than Their Ventilation," ibid. (October 1945), 205; "Where Is the Radiator?" ibid. (July 1945), 21. For an advertisement making identical claims for lighting, see "How Bright Do You Want Your Bright New World?" *Architectural World* 98 (August 1945), 176.

6. Boris Emmet and John E. Jeuck, *Catalogues and Counters: A History of Sears, Roebuck, and Company* (Chicago: University of Chicago Press, 1950), pp. 435–436; "Post-War Status of Domestic Heating Equipment," *Consumers' Research Bulletin* 19 (March 1947), 22.

households sharing space in 1947 and another half million occupying temporary quarters.[7]

So desperate did the housing situation appear that many built or remodeled homes on their own. During the summer of 1946, Jacques Brownson, twenty-three years old, recently married, and a former carpenter's apprentice in the U.S. Army Corps of Engineers, constructed a house in Aurora, Illinois, a Chicago suburb. The editors of *Popular Mechanics* prepared the plans and repaid him for materials, which because of his veteran status they figured he stood "as good a chance as any at getting." Despite a willingness to invest sweat or a lender's capital, however, most young couples lacking carpenter skills, veteran status, and good fortune would have to wait a few more years for building supplies, skilled workers, and stoves and refrigerators to become available to them.[8]

Inexpensive natural gas also remained in short supply. By 1945, pipeline operators and distributors including Henry Doherty had constructed nearly 220,000 miles of natural-gas pipeline across the nation to serve more than 18.6 million gas customers. Yet that pipeline network did not reach cities in the Northeast. Householders in those cities still relied on costlier manufactured gas. According to a survey conducted by the American Gas Association, most residents of northern Delaware purchased manufactured gas for domestic purposes, probably for cooking. Fewer than half used gas for heating hot water, and fewer than 10 percent burned gas for house-heating.[9]

7. "Here Is the Home That Millions Want . . . The *Popular Mechanics* Build-It-Yourself House," *Popular Mechanics* 87 (April 1947), 108; Jackson, *Crabgrass Frontier*, pp. 232–233. For the politics of postwar housing shortages and efforts to reduce them, see Richard O. Davies, *Housing Reform During the Truman Administration* (Columbia: University of Missouri Press, 1966).

8. "Here Is the Home That Millions Want" (quotation on p. 110).

9. American Gas Association Bureau of Statistics, *Survey of Residential Gas Service: Utility Customers, Service Characteristics, Appliance Saturations, by Counties in the United States—October 1949* (New York, 1950), pp. 17, 22. For warnings to consumers about pipeline capacity shortages, see "Post-War Status of Domestic Heating Equipment," p. 23. For explanations of the political economy of natural gas after World War II that facilitated increases in production and distribution capabilities and rising consumption, see John G. Clark, *The Political Economy of World Energy: A Twentieth-Century Perspective* (Chapel Hill: University of North Carolina Press, 1990), p. 115; Gerald D. Nash, *United States Oil Policy, 1890–1964: Business and Government in Twentieth-Century America* (Pittsburgh: University of Pittsburgh Press, 1968), pp. 209–237; and Richard H. K. Vietor, *Energy Policy in America Since 1945: A Study in Business-Government Relations* (New York: Cambridge University Press, 1984), pp. 64–90, 146–162; U.S. Department of Commerce, Bureau of the Census, *Statistical Abstract of the United States, 1960* (Washington, D.C.: Government Printing Office, 1960), p. 536.

Because natural-gas producers held vast reserves, a postwar boom in natural-gas consumption awaited only the installation of additional pipeline mileage, conversion of older buildings from coal to gas, and construction of immense numbers of houses and apartments in new subdivisions. By the mid-1950s, operators had extended lines from fields in the Southwest to large cities in the East, West, and Midwest, including such large markets as Philadelphia, New York, and Boston. During the period 1945–1969, then, the number of residential customers doubled to more than 37 million. By the late 1960s, moreover, most urban Americans not only enjoyed access to natural gas but also used it at a still more rapid pace. Between 1950 and 1969, residential consumption of gas increased nearly 350 percent.[10]

Electrical operators also increased productive capabilities. Several factors peculiar to the management and politics of the electrical industry coalesced around the idea that generating capacity following the war would have to be greatly increased. In part, executives ordered that increase in capacity in anticipation of growing demand for electric service. Richard F. Hirsh points out, however, that the political economy of utility regulation also encouraged managers to favor substantial increases in generating capacity. State regulators permitted utility managers to secure returns only on invested capital rather than on labor and supplies, including fuels. As a result, management had good reason to favor large-scale, capital-intensive technologies that boosted capacity. Between 1950 and 1969, capacity at the nation's private utilities increased from 329 billion kilowatt-hours to 1,329 billion, a jump of more than 400 percent.[11]

Traditions of electrical engineering fixed before World War II also guided executives of the electrical industry as they contemplated growth. In particular, leading executives of an earlier period, such as Doherty and Insull, had built their firms to immense success and prestige based in part on the notion of technological leadership. Although electric (and gas) companies held exclusive franchises in their service territories, argues Hirsh, managers, about two-thirds of whom

10. Vietor, *Energy Policy,* p. 84; Christopher James Castaneda, *Regulated Enterprise: Natural Gas Pipelines and Northeastern Markets, 1938–1954* (Columbus: Ohio State University Press, 1993), pp. 117–118, 140–142, 164–166; Edward W. Constant, "Cause or Consequence: Science, Technology, and Regulatory Change in the Oil Business in Texas, 1930–1975," *Technology and Culture* 30 (April 1989), 437–445; *Statistical Abstract of the United States, 1960,* p. 536; *Statistical Abstract of the United States, 1970,* p. 515.

11. Richard F. Hirsh, *Technology and Transformation in the American Electric Utility Industry* (New York: Cambridge University Press, 1989), p. 81; *Statistical Abstract of the United States, 1970,* p. 507.

had backgrounds in engineering, liked "using advanced forms of technologies that promised to push back previous barriers to improved performance." During the 1950s, for example, executives of the American Gas & Electric Company, serving Ohio and western Virginia, won awards from the General Electric Company for installing high-capacity generating units and high-voltage transmission lines. In 1960, Commonwealth Edison opened the nation's first full-size nuclear power plant using private capital. For about twenty years following World War II, public policy and traditions of engineering practice coalesced around the idea that remarkable increases in electrical capacity were desirable and possible.[12]

Consumption of electricity more than matched expanded output. In the first place, the total number of electric customers increased from 37.5 million in 1950 to 62.5 million in 1969. Consumption of electricity at the residential level during that same period jumped from 67 billion kilowatt-hours to a whopping 408 billion, an increase of more than 600 percent. In turn, declining prices for electric service contributed to this widening and increasing demand. Hirsh provides a valuable example at the local level of these national trends. During the period 1951–1967, he reports, managers of the Virginia Electric & Power Company boosted installed capacity 432 percent, handled an increase in demand of 341 percent, enhanced thermal efficiency 26.5 percent, and managed to cut costs to residential users by 23 percent.[13]

By the early 1950s, home economists and other experts who wrote for a popular audience could urge installation of gas and electric appliances without fear that orders would go unfilled or that utility operators would fail to provide adequate service. In fact, these modern agents of diffusion painted a picture that was quite the reverse. Now it was virtually mandatory, many contended, to install new furnaces, stoves, lighting systems, and kitchen appliances in order to enhance the comfort and convenience of the built environment for family members, especially women. With the proper arrangement of equipment, householders able to finance these purchases could shape a built environment that was also aesthetically more satisfying. Technology, beauty, environmental control, and gender went hand in hand. In reality, popular writers and appliance manufacturers emphasized

12. Hirsh, *Technology and Transformation,* pp. 69–81, including quotation on p. 73.

13. *Statistical Abstract of the United States, 1970,* p. 507; Hirsh, *Technology and Transformation,* pp. 83–84.

Fig. 17. "We're a 3-generation Maytag Family!" 1951. Comfort, convenience, cleanliness, gender, and economy of effort remained the principal themes in postwar advertising of gas and electric appliances. (Courtesy, Maytag Company)

the continuing industrialization of the household and named women as the plant managers.[14]

No room in the household loomed more important for the installation of gas and electric equipment than the kitchen. In June 1955, an expert writing in *Good Housekeeping* asked whether you "smoke up the kitchen when you broil" or "turn out leathery pancakes." A smoke-free room or tastier meals were not matters of upgrading the homemaker's skills. Rather, she "need[ed] a new range." Stoves featured "thermostatically controlled cooking on the top . . . just as you do in the oven." Gas and electric griddles allowed homemakers to "make perfect hamburgers."[15]

As manufacturers increased the size of these new and popular cooking appliances and equipped them with timing devices, home service specialists emphasized volume production in the kitchen. "Old ranges," observed the home equipment editor for *Women's Home Companion* in 1965, "sometimes seem to groan under the load put on them." New equipment, ran the argument, offered special features that permitted homemakers to prepare larger meals. A grill, for example, cooked three "king-size pineapple pancakes" on one side and eight strips of bacon on the other. A Westinghouse range, reported the home service director for the Union Electric Company in St. Louis, could hold "a 20-pound turkey, glazed sweet potatoes, casserole of onions au gratin and a pan of extra dressing." As an alternative, the oven could "bake eight pies at one time." Not only did new ovens hold more food, they were easier to clean. A Kelvinator range, contended the home service director for the electric company in Milwaukee, allowed homemakers to "throw away the battered parts and replace them with clean foil."[16]

14. For the aesthetics of home lighting and interior decoration in the postwar house, see "The New Art of Lighting a House Means *More* Light and *Better* Light Indoors and Out," *House and Garden* 108 (August 1955), 100–101. The best-known and most influential account of the postwar emphasis on domesticity remains that of Betty Friedan, *The Feminine Mystique* (New York: W. W. Norton & Company, 1963). For a perceptive essay contending that Friedan overlooked the popular literature that was urging women to remain active in public affairs, see Joanne Meyerowitz, "Beyond the Feminine Mystique: A Reassessment of Postwar Mass Culture, 1946–1958," *Journal of American History* 79 (March 1993), 1455–1482; and for the relationships developed during the 1950s between the Cold War and domestic consumerism intended for women, see Elaine Tyler May, *Homeward Bound: American Families in the Cold War* (New York: Basic Books, 1988), pp. 162–182.

15. For the quoted materials, see "Are You Cooking Under a Cloud?" *Good Housekeeping* 140 (June 1955), 97. See also "How a Tiny Obsolete 1926 Kitchen Jumped Ahead 30 Years," *House Beautiful* 98 (February 1956), 84–85.

16. "Holiday Cooking Can Be Easier with These New Ranges," *Women's Home Companion* 82 (December 1965), 48–49. For the installation of timing and other control

Comfort and convenience for women extended from kitchen appliances to home heating systems. By the mid-1950s, the design of furnace and duct systems had evolved to the point that writers in popular magazines could speak routinely of heating air to an identical temperature in every portion of the house. The popular name was "perimeter heating," which in practice meant that home builders placed hot-air ducts along outer walls and capped them with diffusers. This system, according to a writer for *Better Homes and Gardens,* had the advantage of spreading warmed air evenly. Perimeter heating eliminated the "cold seventy," which a writer for *American Home* described as "a layer of cold air building up from floor level." Perimeter heating would "scatter and warm" this "cold-air puddle," allowing householders to experience "fewer drafts and less demand for heat." After the war, the idea of uniform heating was so widely accepted that customers associated a lack of uniformity with the onset of illness. In 1958, seventy-nine delegates attending the Women's Conference on Housing "feared drafts cause colds, dry air and dust lead to sinus troubles, and dry air causes dry skin." Participants at the conference, according to this report, "saw a direct connection between good heating and health."[17]

By the late 1960s, home builders regularly included gas-fired furnaces and gas stoves and water heaters in new homes. In 1967, Americans purchased nearly 500,000 new homes. They heated 78 percent of those homes with gas and another 14 percent with electricity, leaving oil with 7 percent of the market. In the chilly North Central region, 94 percent of new homes were heated with gas, a marked increase for a region that two decades earlier lacked pipelines and natural gas. The association of natural gas with household heating and cooking had passed out of the realm of the novel and into the realm of everyday expectation.[18]

If natural gas for house-heating no longer represented a technological and cultural frontier, the association of air conditioning with

---

devices on household appliances, see Florence Stassen, "New Trends in Major Appliances," *Practical Home Economics* 6 (February 1961), 40.

17. Ronald M. Deutsch, "The ABC's of Heating Today's House," *Better Homes and Gardens* 34 (September 1956), 200; "Modernize Your Old Heating System," *American Home* 58 (August 1955), 58; "Floor Registers Score Best in Heating Test," *Science Digest* 40 (August 1956), 76; "What Women Want from Heating System," *Air Conditioning & Refrigeration News* 85 (November 17, 1958), 7.

18. "29% of Single-Family Homes Were Centrally Cooled," *Air Conditioning, Heating & Refrigeration News* 115 (September 30, 1968), 1; Ruth Schwartz Cowan, *More Work for Mother: The Ironies of Household Technology . . .* (New York: Basic Books, 1983), pp. 194–197.

comfort and convenience for family members, and especially women, was still a cultural (and sales) novelty. Inevitably, however, discussions of the benefits of air conditioning replicated arguments about comfort, convenience, cleanliness, and economy of effort that had been advanced by agents of diffusion since the beginning of the century. In July 1960, a writer for the *Saturday Evening Post* reported on the results of a study conducted by a professor of mechanical engineering at the University of Texas. "In the air conditioned homes," ran the report, women had less to do, "because with doors and windows closed, less dust and dirt got into the house." The number of colds decreased, reported the professor, and the amount of time family members spent with one another went up. One housewife contended that a cooled home during the summer encouraged men to dress in more formal attire for parties and that conversations during those get togethers had become "more stimulating." Equally impressive was the proposition that "the rate of pregnancy in the air-conditioned homes showed a significant increase," which was apparently attributable either to "better health [or a] more relaxed home atmosphere." According to the writer, then, "the man who installs air-conditioning equipment is possibly entitled to call himself an 'environmental engineer.'" By 1967, air conditioning appeared in 29 percent of new homes constructed nationwide and in nearly half the homes built in the South.[19]

Promoters of air conditioning for schools also replicated arguments advanced decades earlier for improved classroom heating. After 1910, educators regularly contended that classroom temperature affected deportment and scholarship. In the early 1960s, educators remained convinced that classroom temperatures influenced behavior and learning, transferring their search for that ideal temperature from heating to air conditioning. In January 1961, an architectural engineer reported a study of two school buildings in Florida. The question was whether air conditioning would "affect take-home learning." By 1964, executives of larger school districts had answered the question for themselves. "Heating," two of them argued, "is no longer the central problem." Instead, "ventilation and cooling are the prime considerations in maintaining an optimal thermal environment in a school classroom."[20]

By the late 1960s, with the distribution of air conditioning units

19. John Reese, "The Air-Conditioning Revolution," *Saturday Evening Post* 233 (July 9, 1960), 97; "29% of Single-Family Homes Were Centrally Cooled," p. 1.

20. Henry Wright, "A Definitive Experiment with Air Conditioning," *American School Board Journal* 142 (January 1961), 29; William R. Manning and Lionel R. Olsen, "Air

increasing rapidly, collegiate educators took up the idea of training students for jobs in the emerging field of air conditioning maintenance and repair. By September 1968, administrators and faculty at one of Chicago's junior colleges offered an associate of arts degree for the study of commercial, residential, and industrial applications of air conditioning systems. A trade publication reported that the courses for the degree included "laboratory training in air conditioning." Whether in 1968 and the early stages of air conditioning, or in 1881 as the people of Kansas City stared in amazement at arc lights, abstract and elegant science turned up in such idiomatic forms as that of "laboratory training." Almost ninety years after electricity and the tools it made work appeared in North America, idiomatic science still served as a preferred form of discourse through which ordinary Americans began to envision a magic and mystical potential for what was in reality only another technological system.[21]

By the early 1970s, the fabulous increases in demand for gas and electric service that had characterized the postwar era came to a halt. During the late 1960s, the rate of increase in the demand for electricity had averaged nearly 9 percent a year. For the period 1980–1987, the rate of increase averaged only 1.9 percent, and between 1973 and 1985 demand for natural gas among residential, commercial, and industrial users actually decreased. Instead, the emphasis among householders and business executives was on energy conservation. In practice, energy conservation often meant that many sought to enhance the efficiency of gas and electric consumption and that still others made do in rooms that were darker and colder in the winter and hotter during the summer. Price increases contributed to the growing interest in energy conservation. Between 1973 and 1985, the average price (in 1982 dollars) of electricity for household purposes went up 36 percent, while the price of natural gas increased a whopping 400 percent.[22]

As prices increased and supplies and demand fell, many began to argue that dramatic changes in patterns of energy production and

---

Conditioning: Keystone of Optimal Thermal Environment," *American School Board Journal* 149 (August 1964), 22.

21. "Chicago College Offering Associate of Arts Degree in Air Conditioning," *Air Conditioning, Heating & Refrigeration News* 115 (September 23, 1968), 34.

22. Clark, *The Political Economy of World Energy,* pp. 254–255; Hirsh, *Technology and Transformation,* p. 101; *Statistical Abstract of the United States, 1989,* pp. 562, 566–567. By 1988, the price of natural gas (still in 1982 dollars) had fallen from 203.0 to 125.6 cents, a marked decrease from the high point achieved in 1985 but still nearly double the price level of 1975. See *Statistical Abstract of the United States, 1991,* p. 574.

consumption were under way. On April 18, 1977, President Jimmy Carter described a program for dealing with price increases and falling supplies as "the moral equivalent of war." By the mid-1980s, rearrangements in the energy industries appeared so dramatic that historian Martin V. Melosi identified an energy transition comparable to the shift from coal to petroleum and gas that began at the turn of the century.[23]

Fundamental changes in several of the political, organizational, and technological dimensions of the gas industry helped bring about these stunning increases in rates. Between 1890 and 1940, public policy at the local and state levels had forced gas utility executives to accept lower rates in return for political peace and long-term financial opportunity. Rate politics, and especially the rate politics of higher-income residents located far from downtown, had encouraged such innovative executives as Henry Doherty to develop formulas for delivering lower prices in the form of mass sales to citywide and then regional markets. Between 1979 and the mid-1980s, however, deregulation of gas rates failed to bring forth additional supplies, leading only to rising prices, falling demand, and uncertainty regarding service. According to business historian Richard H. K. Vietor, "policies designed to deal with shortages and rising prices failed miserably."[24]

Although the political economy and technology of electricity in the postwar years were different from those of gas, the net result was about the same. By the 1970s, the electric utility business had reached a state characterized by Richard Hirsh as one of "technological stasis." No longer could electrical engineers achieve quick and inexpensive increases in the productivity of generating plants. New and highly efficient plants also required longer periods of maintenance, reducing what engineers called the "availability factor." Worse yet, argues Hirsh, engineer-managers of the 1970s and 1980s "lacked the imagination and critical faculties of their professional ancestors of the early twentieth century." Many of these engineers were unable to comprehend that technological limits had been reached. Nor, adds Hirsh, did they recognize that the rapid growth in demand characteristic of the 1920s

23. Martin V. Melosi, "The Third Energy Transition: Origins and Environmental Implications," in Robert H. Bremner, Gary W. Reichard, and Richard J. Hopkins, eds., *American Choices: Social Dilemmas and Public Policy Since 1960* (Columbus: Ohio State University Press, 1986), p. 187 (Carter quotation on p. 199).

24. Clark, *The Political Economy of World Energy,* pp. 254–255; Richard H. K. Vietor, "Government Policy and Energy Markets: A Review Essay," *Journal of Policy History* 5, no. 4 (1993), 274.

Fig. 18. Kitchen of the 1980s. Designers of kitchens still arranged appliances according to the logic of assembly-line production. During the postwar years, however, new machines, such as the microwave and toaster oven, permitted more specialized applications of gas and electricity to the unceasing search for cleanliness, comfort, convenience, and economy of effort. Decorative elements such as the area rug, and the emphasis on color, spoke in a silent fashion to the feminization and domestication of this "factory." (Courtesy, Maytag Company)

and the two decades following World War II had slowed. Unlike earlier executives such as Samuel Insull and Henry Doherty, then, this new generation of electric (and gas) managers could not adapt their organizations and technologies to rapid changes in their regional economies or to equally rapid changes in the politics of energy production and consumption.[25]

25. Hirsh, *Technology and Transformation,* pp. 89–91, 93, 111, 130. See also Douglas D. Anderson, *Regulatory Politics and Electric Utilities: A Case Study in Political Economy* (Boston: Auburn House Publishing, 1981); and Louis C. Hunter and Lynwood Bryant, *A History of Industrial Power in the United States, 1780–1930,* vol. 3: *The Transmission of Power* (Cambridge: MIT Press, 1991), p. 243. During the 1970s and early 1980s, many

Among many who advised Americans on heating, cooling, cooking, and lighting, the response to energy shortages and rising prices was to advocate increased efficiency. Despite substantial changes in the politics, production, and economics of gas and electric service, popularizers and experts advanced the idea that, with enhanced efficiency, no one need sacrifice comfort and convenience. Instead, Americans had only to insulate walls and attics, install fans, and purchase more energy-efficient equipment. Writers in such magazines as *Glamour* and *House & Garden* advised their mostly female readership to "Look for Energy Guide Labels When You Shop" and to "Read the Label—Save Energy." One writer contended: "The new range of tools for easier food preparation includes as many electrical gadgets as before, but . . . the best and brightest of them are being designed for maximum energy efficiency."[26]

As part of the emphasis on energy efficiency, experts in newly emerging fields, such as environmental analysis and energy management, prepared reports on topics like methods for reducing energy costs without a loss in productivity, safety, or comfort. Investments in lighting improvements, the reasoning went, could even lead to cost reductions in related areas. Not only did a high-efficiency system for outdoor lighting at Central Michigan University cost less to operate than the old system, an engineer reported in 1985, but school officials had also determined that they could reduce the expense of having security patrols.[27]

---

householders installed wood and even coal stoves in kitchens and basements, no doubt hoping that possession of a wood stove would allow them to express a revived interest in the natural and in small-scale technologies. Just as important, however, was the low price of wood compared with electricity and even gas for heating. By January 1980, *Popular Science* was able to describe the advantages of nine stoves, including one unit the manufacturer claimed was able to heat a family's hot water and most of its house. See "Buyers Guide to Coal/Wood Stoves," *Popular Science* 2 (January 1980), 123.

26. Larry Stains, "Understandable Energy Tips," *Family Handyman* 35 (July–August 1985), 102, 105; "Help Your House Beat the Heat," *Better Homes and Gardens* 63 (July 1985), 125; "Look for Energy Guide Labels When You Shop," *Glamour* 78 (October 1980), 50; "Read the Label—Save Energy," *House and Garden* 152 (May 1980), 86; Duncan H. Maginnis, "How to Keep Your Cooking Up-to-Date," *House and Garden* 153 (April 1981), 74.

27. James C. King, "Lighting Boosts Safety, Security, Production," *National Safety and Health News* 132 (November 5, 1985), 89–91. For an example of continuing interest in the relationship between behavior and lighting as part of the built environment extending back to the period before World War I, see Sheila Danko, "Lighting Research and Application: Are We Still in the Dark?" *Human Ecology Forum* 15, no. 4 (1986), 13–16. For an example of the logic of research in the field of energy conservation with implications for shaping public policy, see Lester W. Baxter et al., "An Efficiency Analysis of Household Energy Use," *Energy Economics* 8 (April 1986), 62–73.

Still others overlooked rising prices and energy shortages. Women's magazines continued to focus on methods to enhance the taste of foods and to ease the burdens of cooking. In February 1982, *Better Homes and Gardens* published an article entitled "Classic Cooking Techniques to Make Simple Fare Sensational." The following month, *Redbook* was advising readers, "At Last! How to Snap out of a Cooking Rut." In December 1986, economist and journalist Jane Keely highlighted new kitchen appliances such as food processors, electric pasta makers, ice-cream freezers, and crepe makers. In particular, she liked the "mini-model" food processor for specialized tasks like "mincing, chopping and grinding about a cupful of onions, bread crumbs, cheese, coffee beans, peppercorns, nutmeg, and the like." Although many cooks already owned a full-size food processor, she added, they "might well want a . . . mini-model processor for quick, simple preparation of single recipe ingredients."[28]

Between 1880 and the late 1960s, then, Americans had redefined gas and electric service away from the realm of magic and mysterious science intended for the few, and into the realm of technology, which was judged accessible to most. By the mid-1960s, those with at least modest incomes had embedded the light and heat brought by immense gas and electric systems into their schools, homes, and places of worship. After 1970, rising prices and declining service signaled a change in the political economies of gas and electric production. Despite substantial price increases, however, the idea that cleanliness, comfort, and convenience were contingent on abundant supplies of gas and electricity remained one of the fixtures of American popular culture and domestic practice, particularly among women. Patterns of gas and electric diffusion and utilization established between 1880 and 1940 by teachers, home builders, architects, and other agents of diffusion persisted at least up to the last decade of the century in the habits of millions of Americans and in the advice presented by still another generation of agents of technological and social change.

28. J. Taylor, "Classic Cooking Techniques to Make Simple Fare Sensational," *Better Homes and Gardens* 60 (February 1982), 98; "At Last! How to Snap Out of a Cooking Rut," *Redbook* 158 (March 1982), 72; Jane Keely, "What's Cooking with Kitchen Electronics," *Consumer Research Magazine* 69 (December 1986), 17–18. For the useful observation that postwar ideologies regarding the appropriate roles of women were "varied . . . and complex," see Meyerowitz, "Beyond the Feminine Mystique," p. 1480.

Epilogue: The Scholarship of Technology and Society, 1915–1990s

Just as Marx posited that the economic system evolved in a manner almost independent of human will, to [William F.] Ogburn the technological system was largely independent of the thoughts and actions of individuals.

—Sociologist Ron Westrum, 1991

Like other students of the social construction of technology, I believe that it is a mistake to separate the artifact from its organizational setting.

—Historian W. Bernard Carlson, 1992

The diffusion question was neglected.

—Geographer Lawrence A. Brown, 1981

THE GROWING POPULARITY OF GAS AND ELECTRIC appliances after World War II coincided with a process among scholars of reexamining the causes and consequences of technological and social change. Beginning in the early 1920s, William F. Ogburn and other sociologists at the University of Chicago published studies describing industrial and urban impacts upon residents of cities. Ogburn in particular described technological changes that were taking place at such a rapid pace that urban residents often lagged in their response. After World War II, however, the intellectual authority of Ogburn and other Chicago sociologists receded. Members of subsequent generations of scholars inverted many of the findings of Chicago sociology. By the mid-1980s, a group calling themselves social constructivists were contending that social and economic factors determined technological outcomes—not the other way around, as Ogburn had argued.

This book has been grounded in the work of contemporary scholars articulating the idea that society determines its own technological outcomes. Equally, a large body of literature by historians and sociologists that extends back to the University of Chicago during the 1920s and 1930s has informed my understanding of rapid social and industrial change in cities. Yet authors of both of those rich and detailed bodies of publications, I learned, failed to explain several of the developments on the gas and electric scenes in Kansas City and Denver before World War II or throughout the United States during the postwar years.

Earlier scholars of technology and society emphasized the infinite expandability of electrical and gas systems rather than the limits on their growth. Earlier scholars, I also learned, neglected the rapid growth and the political economy of cities in determining the marketing strategies and prices of gas and electric operators after 1880 and extending until the end of the next century. Nor did those scholars, including the social constructivists, perceive the role played by agents of technological diffusion, including educators and numerous salespersons like Roy Munroe and J. C. Nichols. Agents of diffusion such as Munroe and Nichols defined gas and electric appliances in terms of hygiene, comfort, and convenience, and brought that message to the doorstep of every urban householder and into every classroom in the nation.

The place to start this reappraisal of the evolving scholarship of technology and society is with the agreed-on elements of an urbanizing and industrializing nation. Beginning around 1880, gas, electricity, light, and heat comprised only one portion of a much larger picture of technological innovation and systems-building taking place in virtually every city. Between 1880 and 1930, large factories grew up alongside small workshops; electric trolley systems employing hundreds and then thousands covered the city with tracks, replacing a jumble of steam railroads, cable cars, and horse-drawn vehicles; and the city itself increased in size from a few miles and perhaps a few thousand persons to a region 40 or 50 miles in diameter and serving as home and workplace to several hundred thousand or perhaps several million people. In turn, residents of large cities and smaller ones clustered their places of residence according to race, ethnicity, and income. Finally, politicians, business executives, and experts in finance, engineering, and administration competed with one another for the opportunity to shape the landscape and to govern these technological systems and the metropolis.[1]

Scholars have sought to understand these remarkable transformations in the size of cities and in the relationships of urban-dwellers with one another by developing concepts of social, urban, and technological change. In the United States, serious study of the relationships between industrial and urban change actually began at the U.S. Census Office, which later became the Bureau of the Census. In 1886, the census office published two volumes detailing the social statistics of cities. Prepared by George E. Waring Jr., these volumes described important and diverse components of the built environment, such as highway paving, gas works, and street railroads, in more than 200 cities. Although his account of developments in these cities lacked an overarching theme, Waring nonetheless regularly associated modern technological systems with the health and comfort of millions.[2]

In 1899, economist Adna Ferrin Weber published a smaller and more focused account of the rapid growth of cities throughout the world. Transportation and demographic trends were at the center of Weber's study. On the one hand, he thought, the trolley was more a response to "the American *penchant*" for suburban living than its cause. On the other, he perceived the "transcendent importance of rapid transit as a remedy for overcrowding."[3]

More important in the long run for scholarly and popular understanding of urban growth and change than any of the detailed find-

1. Thomas P. Hughes, *Networks of Power: Electrification in Western Society, 1880–1930* (Baltimore: Johns Hopkins University Press, 1983), pp. 201–226; Harold L. Platt, *The Electric City: Energy and the Growth of the Chicago Area, 1880–1930* (Chicago: University of Chicago Press, 1991), pp. 3–21; Paul Barret, *The Automobile and Urban Transit: The Formation of Public Policy in Chicago, 1900–1930* (Philadelphia: Temple University Press, 1983), pp. 9–45; Charles W. Cheape, *Moving the Masses: Urban Public Transit . . . 1880–1912* (Cambridge: Harvard University Press, 1980), pp. 1–39.

2. U.S. Department of the Interior, Census Office, *Report on the Social Statistics of Cities, Part I: The New England and the Middle States* (Washington, D.C.: Government Printing Office, 1886; New York: Arno Press, 1970). For a sustained analysis of the data and observations contained in this volume and its companion, see Lawrence H. Larsen, *The Urban West at the End of the Frontier* (Lawrence: Regents Press of Kansas, 1978).

3. Adna Ferrin Weber, *The Growth of Cities in the Nineteenth Century: A Study in Statistics* (New York: Macmillan Company for Columbia University, 1899; Ithaca: Cornell University Press, 1963), pp. 197–199, 469–473 (quotations on pp. 469, 471, with emphasis in the original). At the meeting of the Urban History Seminar of the Chicago Historical Society held on April 22, 1992, Michael H. Ebner described Weber's book as the "Old Testament" of urbanistic scholarship. I want to thank Professor Ebner for sharing that insight with me.

ings of Waring or Weber were their asserted and implied statements about the influence of transport, industry, and population movements in shaping the direction of change among urban residents. Indeed, their work gave the weight of scholarly and statistical authority to the notion that large industry and vast technological systems had played a substantial part in determining the land uses, work, and domestic lives of urban residents.

Beginning in 1915 and continuing up to World War II, sociologists located at the University of Chicago launched the first sustained analysis of technology and society in the American city. "Chicago," argues historian William H. Wilson, "was at the center of things sociological." Several scholars, described by Wilson as "hyperactive," dominated the university's Department of Sociology. Within that group, five members of the faculty, including Ernest W. Burgess, Roderick D. McKenzie, William F. Ogburn, Robert E. Park, and Louis Wirth, often brought the relationship between industry, technology, and the city front and center.[4]

Everywhere in urbanizing and industrializing America, Chicago sociologists perceived the tangible signs of social and physical change, including large immigrant populations, growing corporations, and metropolitan areas spread across vast landscapes. Although members of the "Chicago school" and their graduate students completed an impressive number of empirical studies during the interwar years, focusing on immigrants, industry, transportation, communication, geographic dispersion, and a variety of social problems, some general propositions guided their research. Several at Chicago stressed the consequences of vast population movements and cultural mixing; others emphasized the rational and impersonal calculations of mass production, distribution, and consumption. All were agreed, however, that the introduction of these forces was leading to the collapse of traditional and intimate ties of family and town, church and synagogue. As early as 1916, Robert Park had concluded that in the city

4. William H. Wilson, *Coming of Age: Urban America, 1915–1945* (New York: John Wiley & Sons, 1974), pp. 92–118, and quotation on pp. 92–93. Studies by University of Chicago sociologists were part of a U.S. and European scholarly tradition of seeking to comprehend the origins of the modern era. For early examples of the European tradition of modernization studies, see Ferdinand Tonnies, *Community and Society,* trans. and ed. Charles P. Loomis (New York: Harper & Row, 1963; originally published in 1887), where the emphasis was on cultural change, especially rationalization; and Emile Durkheim, *The Division of Labor in Society,* trans. George Simpson (Glencoe, Ill.: The Free Press, 1966; originally published in 1893), where the emphasis was on economic restructuring, especially the division of labor. Claude S. Fischer helped me clarify these points.

the "intimate relationships of the primary group are weakened and the moral order which rested upon them is gradually dissolved."[5]

One of these Chicago sociologists, William Ogburn, contended that technological systems and mechanical devices, rather than population movements or the growth of large cities, were at the center of social change. Because of the fast growth of technological systems, Ogburn believed that the process by which householders or business leaders adapted to them was slow and awkward. Recent inventions such as steam or electricity, Ogburn asserted in 1922, were "adopted with the idea of satisfying individual wants, because they bring comfort, rest, speed, enlightenment, or wealth." For all their advantages, however, these technologies demanded adjustments among members of society that "may require a considerable number of years, during which there may be said to be a maladjustment." This slowness to adapt to technology, Ogburn contended, was the result of a process of cultural lag. By 1957, two years before his death, Ogburn had reduced the relationship between technology and culture to an exact form. "Lags accumulate," he believed, "because of the great rapidity and volume of technological change."[6]

During the early 1930s, Ogburn's idea left the ivory tower. In 1933, leaders of the world's fair held in Chicago popularized the notion of technology's sovereignty with the motto "Science Finds—Industry Applies—Man Conforms." Perhaps, as historian Carroll Pursell argues, this motto spoke to a popular conviction that technology "was continuing to produce great changes in American society." Regardless

5. Robert E. Park, "The City: Suggestions for the Investigation of Human Behavior in the Urban Environment," reprinted in Richard Sennett, ed., *Classic Essays on the Culture of Cities* (New York: Appleton-Century-Crofts, 1969), p. 130; R. D. McKenzie, "The Rise of Metropolitan Communities," in President's Research Committee on Recent Social Trends, ed., *Recent Social Trends in the United States,* vol. 1 (New York: McGraw-Hill, 1933), p. 444. See also Richard Wightman Fox, "Epitaph for Middletown: Robert S. Lynd and the Analysis of Consumer Culture," in Richard Wightman Fox and T. J. Jackson Lears, eds., *The Culture of Consumption: Critical Essays in American History, 1880–1980* (New York: Pantheon Books, 1983).

6. William F. Ogburn, *Social Change with Respect to Cultural and Original Nature* (New York: Dell Publishing Company, 1966; originally published by B. W. Huebsch, 1922), pp. 200–213, 268–278, and quotations on pp. 201 and 270; idem., "Cultural Lag as Theory," in Otis Dudley Duncan, ed., *William F. Ogburn on Culture and Social Change: Selected Papers* (Chicago: University of Chicago Press, 1964), p. 92. Ogburn and his colleagues found no need to study the processes by which these new technologies actually entered into the daily lives of urban residents. According to geographer Lawrence A. Brown, *Innovation Diffusion: A New Perspective* (New York: Methuen, 1981), p. 179, Chicago sociologists "took as a given that the invention would find its way into the economy due to its superiority over the old technology."

of the reason, during the 1930s Ogburn's contention that technology had taken the lead and that society was lagging behind moved from the esoteric arena of scholarly books and articles into the realm of corporate advertising and public relations.[7]

By the late 1930s, the intellectual primacy of Chicago sociology was receding. After World War II, scholars of technology and society turned their attention in two directions. First, they emphasized preparation of studies that were more limited in scope, dropping such all-encompassing generalizations as cultural lag. Writing from the point of view of the mid-1980s, historian Fred H. Matthews found "Chicago studies of ethnic communities . . . to be triumphs of large-scale conceptualization over the particularities of lived experience." Second, postwar scholars of technology and society began to promote the idea that social institutions, such as religion, neighborhood, family, class, gender, and work, determined the meaning attached to each new device. Rather than affirming the omnipotence of technique and technology, members of subsequent generations of scholars contended, for example, that many new technologies actually permitted people to do what they had always done, but in another way. A study of telephone users during the period 1900–1940 makes this point. According to sociologist Claude S. Fischer, "Americans apparently used home telephones to widen and deepen existing social patterns rather than to alter them." Since the 1950s, scholars of society and technology have been increasingly disposed to the idea that social and economic institutions have exercised a predominant hand in shaping the meanings and uses of mechanical devices and technological systems.[8]

7. Carroll W. Pursell Jr., "Government and Technology in the Great Depression," *Technology and Culture* 20 (January 1979), 162, brought the fair and its motto to my attention. See also Dorothy Ross, *The Origins of American Social Science* (New York: Cambridge University Press, 1991), p. 444, for the observation, consistent with Pursell's, that Ogburn's notion of cultural lag "gave a name and shape to the widespread sense of historical malaise and it expressed in seemingly scientific form the new liberal hope for progress by incremental change." Equally, for the contention that designers of exhibits at the New York World's Fair of 1964 represented technology as a "black box" that was inexorably shaping human relationships, see Michael L. Smith, "Making Time: Representations of Technology at the 1964 World's Fair," in Richard Wightman Fox and T. J. Jackson Lears, eds., *The Power of Culture: Critical Essays in American History* (Chicago: University of Chicago Press, 1993), pp. 228–229.

8. Fred H. Matthews, "Louis Wirth and American Ethnic Studies: The Worldview of Enlightened Assimilationism, 1925–1950," in Moses Rischin, ed., *The Jews of North America* (Detroit: Wayne State University Press), p. 126; Claude S. Fischer, *America Calling: A Social History of the Telephone to 1940* (Berkeley and Los Angeles: University of California Press, 1992). For an account of historiographical innovations since World War II in the area of technology, cites, and business leadership, see Charles N. Glaab,

Yet contemporary scholars have not been satisfied merely to assert the sovereignty of society. Instead, many are convinced that between a society and its technological systems the relationships are so subtle and contextually dependent that we must "unpack" each technology (for example, the telephone or kitchen appliances) and seek its origins and meanings amid its own group of inventors, investors, and users. In this scheme of things, the meaning and use of these everyday machines are matters of interpretation and negotiation, often across a contested terrain.

Take the cases of the telephone and domestic appliances. Before 1900, argues Fischer, executives at the American Telephone & Telegraph Company anticipated that the principal market for their device would be business leaders like themselves. Although telephone com-

---

Mark H. Rose, and William H. Wilson, "The History of Kansas City Projects and the Origins of American Urban History," *Journal of Urban History* 18 (August 1992), 371–394. See also Brian J. L. Berry, *The Human Consequences of Urbanization: Divergent Paths in the Urban Experience of the Twentieth Century* (New York: St. Martin's Press, 1973), pp. 1–66, for a critique of the findings of the Chicago school. Berry contends that the contemporary population is composed of subcultures and specialists who, though widely dispersed across the urban landscape, employ electronic communications to maintain communities of interest. On the other hand, many scholars continue to affirm the singular importance of the trolley, and then the highway, in fostering social and urban change. As an example, Kenneth T. Jackson, *Crabgrass Frontier: The Suburbanization of the United States* (New York: Oxford University Press, 1985), p. 103, identifies a broad range of social and political factors in fostering the process of suburbanization, but he assigns a unique role to the trolley. "No invention," he argues, "had a greater impact on the American city between the Civil War and World War I than the visible and noisy streetcar and the tracks that snaked down the broad avenues into underdeveloped land." A minority tradition in the study of technology and society stressing the preeminence of technology in everyday affairs, and especially its pronounced effects on the economy, social relationships, and personality continues to win strong sales and positive reviews. Fischer, *America Calling*, p. 12, finds that authors of these books impose a technological determinism that is "'soft,' complex, and psycho-cultural" as opposed to the older tradition of technological determinism deriving from the Chicago school, which he describes as "'hard,' simple, and mechanistic." For a popular example of this more contemporary and "soft" form of technological determinism, see Alvin Toffler, *Future Shock* (New York: Random House, 1970), and for a sophisticated and perceptive argument that dependence on technological systems renders them our masters, see Langdon Winner, *Autonomous Technology: Technics-out-of-Control as a Theme in Political Thought* (Cambridge: MIT Press, 1977). Finally, see Thomas J. Misa, "How Machines Make History, and How Historians (and Others) Help Them to Do So," *Science, Technology, and Human Values* 13 (Summer and Autumn 1988), 308–323, for the argument that scholars who study technology from a "macro" perspective, such as philosophers, are more likely to find technology determinative than others, such as labor historians, who work at the "micro" or shop-floor level of analysis. I want to thank Professor Misa for sending a copy of this article to me.

pany executives directed their early promotional activities to corporate executives and business uses, householders—especially women—defined the telephone as a mechanism for enhancing social relationships. Eventually, telephone company executives reoriented marketing campaigns to take this household-level determination of a technology's uses into account.[9]

Ruth Schwartz Cowan advances a similar argument regarding the purchasers of domestic appliances. During the 1950s, prosperous householders bought new appliances—stoves, refrigerators, washing machines, vacuum cleaners. Cowan finds that these householders took on a number of tasks with their new appliances that could have been performed by machines in factories, including baking, knitting, and hand-dipping chocolates. In turn, Cowan identifies a series of "historical fantasies" during this decade about the appropriate roles of men and women that led to a "backward search for femininity." Altogether, Cowan and Fischer bring the householder back into the process of defining and supervising technologies, and they do so in a fashion that specifies the significance of a particular machine for the gender, income, habits, popular ideas, and contemporary ideologies of buyers and users during a specific period of time.[10]

9. Fischer, *America Calling,* pp. 60–85. See also Claude S. Fischer, "'Touch Someone': The Telephone Industry Discovers Sociability," *Technology and Culture* 29 (January 1988), 32–61. On the concept of unpacking technologies, see Claude S. Fischer and Glenn R. Carroll, "Telephone and Automobile Diffusion in the United States, 1902–1937," *American Journal of Sociology* 93 (March 1988), 1173–1174.

10. Ruth Schwartz Cowan, *More Work for Mother . . .* (New York: Basic Books, 1983), pp. 205–206. For clear statements of the contingent and negotiated character of technological devices, see John Law, "Technology and Heterogeneous Engineering: The Case of Portuguese Expansion," in Wiebe E. Bijker, Thomas P. Hughes, and Trevor Pinch, eds., *The Social Construction of Technological Systems . . .* (Cambridge: MIT Press, 1987), pp. 111–132. During the 1970s and 1980s, urban historians published a large body of literature detailing the ways in the which politics, the economy, or a group of persons such as alley dwellers directed the alteration of technological systems in cities. See, for example, James Borchert, *Alley Life in Washington: Family, Community, Religion, and Folklife in the City, 1850–1970* (Urbana: University of Illinois Press, 1980); Barrett, *The Automobile and Urban Transit;* Christine Meisner Rosen, *The Limits of Power: Great Fires and the Process of Growth in America* (New York: Cambridge University Press, 1986); and Lizabeth Cohen, *Making a New Deal: Industrial Workers in Chicago, 1919–1939* (New York: Cambridge University Press, 1990), pp. 133–138. As one should expect, the shift among scholars of technology and society from the universal to the specific and contingent has encouraged a reconsideration of the theoretical and methodological underpinnings of the field. See, for example, Philip Scranton, "Theory and Narrative in the History of Technology: Comment," *Technology and Culture* 32 (April 1991), 391, for his assertion, consistent with this argument, that "the doing of both history and social science no longer can be linked to the establishment of transcendent verities or the validation of

During the 1980s, historians of technology also began the process of unpacking their analyses of the growth and operation of electrical systems constructed in large cities during the period between 1880 and the 1930s. Thomas P. Hughes studied the inventors, engineers, and investors who built the great utility systems in London, Berlin, and Chicago. Politics mattered greatly in shaping local electric systems, and Hughes found that political leaders in each city entertained radically different ideas regarding the degree of regulation that city officials should exercise, and even differed as to the appropriate scale of electrical operations. Politicians in London stressed "the primacy of politics," argues Hughes, leaving the city in 1914 with "sixty-five electrical utilities, seventy generating stations . . . , and about seventy methods of charging and pricing." By contrast, Hughes describes Chicago's politicians as "venal" and "pliable," which meant that, in practice, the manager-entrepreneur Samuel Insull was able to "relegat[e] political authority to the subservient position deemed appropriate by American entrepreneurs." By 1910, in the absence of a strong regulatory hand, Insull offered electrical service throughout the Chicago region at low prices, and contemporaries judged his system "to be the world's greatest." In Chicago, then, Hughes finds that electrical service was "relatively free of conservative, or traditional, constraints." Whether in London or Chicago, urban context and the contingencies of urban politics determined the organization of electrical systems and the price of electrical service.[11]

By the late 1980s, Hughes and others were increasingly identifying their work with that of a small but influential group of historians and sociologists who called themselves social constructivists. Briefly, social constructivists put social and political considerations first in determining the pace and direction of technological change. Members of groups, went the reasoning, determined the meaning of machines. Naturally, the meaning and utility of a machine varied according to time and place. "An artifact," according to Wiebe E. Bijker, "is gradu-

---

inexorable processes." Instead, he urges attention to "midrange theories that seek to establish the sources of durable patterns rather than the basic principles of human action." Finally, see W. Bernard Carlson and Michael E. Gorman, "Understanding Invention as a Cognitive Process: The Case of Thomas Edison and Early Motion Pictures," *Social Studies of Science* 20 (1990), 388, for their sensible contention that scholars of the technology-society relationship must seek not only to explain the context of each technology but also to explain the cognitive structuring of each inventor.

11. Hughes, *Networks of Power,* pp. 202, 204, 208, 227, 260–261. See also Platt, *Electric City.*

ally constructed or deconstructed in the social interactions of relevant social groups."[12]

During the course of the twentieth century, then, two traditions have guided scholarly and popular understanding of urban and industrial change. Proponents of an older tradition launched by Weber and Waring and systematized during the 1920s and 1930s by Chicago sociologist William Ogburn assigned technology the primary role in determining the pace and direction of social change. Because Ogburn awarded a virtually autonomous life to technology, it followed in his mind that technology would change first and society next, inevitably creating a cultural lag. By the 1930s, popularizers had reduced his studies to the crude but memorable notion that once industry had determined a technological process, "man conforms."

Proponents of a social history of technology have argued the matter quite in reverse. Social historians have been quick to assert the primacy of society, economy, and politics in governing the shape and pace of technological change. By the 1980s, moreover, scholars of technology and society endorsed the idea of "unpacking" technologies. Words like "contingent" and "contextual" loom large among these scholars, informing studies identifying networks of actors (including their gender) who negotiated the meaning of such artifacts as the vacuum cleaner, the telephone, or the gas stove.

Such is the state of the art at one juncture of the history of technology, cities, and social change. Patterns of gas and electric development in Kansas City and Denver up to 1940 and in the United States as a whole after 1945 suggest that we may proceed a bit further. In particular, this study highlights the importance of several contexts in determining the organization of gas and electric firms, their rates, their distribution systems, and the meaning that ordinary Americans attached to new appliances.

Cities like Denver and Kansas City comprised one of those contexts. Before 1900, people with many years of experience in town building sought gas and electric franchises as another outlet for their energies, capital, and connections to politicians. The first gas and electric opera-

12. Wiebe E. Bijker, "Do Not Despair: There Is Life After Constructivism," *Science, Technology, and Human Values* 18 (Winter 1993), 119. Most historians would not classify themselves as constructivists. Nonetheless, social construction receives considerable attention in a few scholarly journals and among editors at several university presses. The situation appears similar to one described by Michael Frisch for the state of urban history, in which "social-scientific invocation controlled the press box, and the idiographic tradition ruled the playing field." See Michael Frisch, "American Urban History as an Example of Recent Historiography," *History and Theory* 18 (1979), 356.

tors treated their new firms as extensions of earlier and fabulously lucrative activities such as securing the franchise to operate the city water supply or securing a railroad connection near property for which they held an option.

Cities were also an unstable context. By the mid-1890s, those committed to the idea of profiting from rapid growth under the aegis of a city franchise were overtaken by that growth and by the politicians on whom they had relied. Virtually all agreed that rates were too high. Even more, many of the same political leaders joined with residents of outlying areas to demand extensions of gas and electric service miles from downtown and at rates no higher than rates charged residents who were much nearer plants. By 1900, most of the original operators had vacated the field in favor of outside capital and specialized teams of managers with years of experience.

Capital and expertise alone were insufficient. Only those who adapted to the fast-expanding geography and changing politics of their cities survived. In Kansas City, not even the economies derived by linking electric and trolley service were enough to rescue Armour's electric company from bankruptcy. Equally, a history of inadequate gas service and rates that appeared too high encouraged politicians to insist on extraordinarily low rates from executives of the old company who were seeking the city's natural gas franchise. Bankruptcy followed there too.

Henry Doherty was one of the first utility executives to align rates and distribution systems with the growth and political geography of his city. He made rates uniform throughout the city. Doherty's rates were also lower for those who used more gas and electricity. In all, he structured distribution networks and prices in a fashion that aimed to satisfy wealthier householders located in new subdivisions southeast and east of downtown.

After 1900, savvy executives like Doherty responded to lower rates and mandated service areas (and high fixed costs) by launching massive sales campaigns. Doherty created a large sales force and directed them to canvass every householder. Experience and their own preferences taught members of that sales force that appeals to cleanliness, comfort, convenience, and economy of effort were most effective in selling irons or gas water heaters, especially in presentations to women. Public policy had encouraged Doherty to set in motion a program of mass production and mass consumption of light and heat that placed gender at the center of the sales pitch.

Members of Doherty's sales force served as agents of technological diffusion. After 1900, teachers, architects, and countless others joined them in informing Americans about the advantages of gas and electric service. For example, educators linked bright lights and improved ventilation to campaigns already under way to improve personal hygiene and public health.

During the 1920s, agents of diffusion such as J. C. Nichols and Roy G. Munroe brought routine form to the sale and use of gas and electricity. Nichols, developer of Kansas City's Country Club District and Plaza, stood between a generalized desire for comfort and convenience, on the one side, and complex and little understood infrastructural systems, on the other. Munroe, a salesperson and later a sales manager, took charge of bringing messages about the comfort and convenience of new appliances to every householder in Denver. For both Munroe and Nichols, the obligation of men to protect women from the hazards of industrializing cities and the burdens of household management occupied a prominent place in every message. Again, gender mediated the diffusion process. By 1940, then, the city and countless agents of diffusion had created contexts for the vast popularity of gas and electric appliances that took place after the war. Those contexts had also suggested the very meaning of those appliances in the domestic arrangements and gender relationships of the next generation.

Between 1945 and the late 1960s, Americans purchased vast numbers of appliances. Gas and electric consumption increased rapidly. Whether in the form of classroom instruction or in advertisements, the point agents of diffusion stressed after 1945 was the responsibility of men for purchasing larger pieces of domestic equipment, such as air conditioning, and the continuing responsibility of women for maintaining a comfortable home and preparing large and interesting meals. In turn, the availability of heated baptismal tanks and air-conditioned offices located near air-conditioned automobiles and trains made it increasingly possible for higher-income Americans to spend large portions of their lives in controlled environments. As part of their celebration of this prosperous, comfortable, and energy-dependent society, scholars and popular writers spoke knowledgeably about an "American character," and even of an "American civilization."

The reality was more prosaic. Following the war, institutionalized behaviors and political arrangements had converged, encouraging Americans to build environments of light and heat that were spread across an immense landscape. Federal and state governments built highway networks that allowed Americans to live far from central cities. Government also financed construction of water and sewer systems extending into distant suburbs, all the while guaranteeing the mortgages of new residents. At the same time, electric and gas rates declined; engineers built larger and more efficient plants; regulators kept energy prices low, particularly prices for natural gas; and lengthy pipelines and electrical interchanges carried that energy throughout the continent.

After 1970, these arrangements began to uncouple. First, engineers discovered that their ability to drive unit costs lower was limited. Shortages of natural gas added to inflationary pressures in the energy industries and in the budgets of most Americans. In response, politicians and economists came to believe that deregulation would lead to lower prices and increased production. Up to the mid-1980s, however, the consequences were vastly increased prices, continuing shortages, and rooms that were hotter in summer and chillier in winter. About the only persistent item on the gas and electric scene was the continuing insistence of agents of diffusion that gas and electric appliances would enhance cleanliness, comfort, and convenience, especially for women.

The meanings, uses, and patterns of diffusion that characterized gas and electric service and light and heat were negotiated within the contexts of several periods of urban and national history. Gas and electric service and built environments of comfort and convenience were constructed in the changing political economy and geography of rapidly growing cities, in elite and mass politics, in public schools teaching hygiene and job skills for men and women, in the design of homes for the well-off, in employee training programs, and in the ceaseless efforts of such agents of diffusion as J. C. Nichols and Roy Munroe. The political and other institutions that had facilitated the widespread adoption of gas and electric systems were renegotiated during the period around 1900, during the 1920s and 1930s, again after 1945, and once again during the period between the late 1960s and the early 1980s. Only the results of those agents of diffusion in shaping the habits and outlooks of ordinary Americans regarding the meaning of light and heat persisted up to the present day. On the other hand, ideas about hygiene or comfort and the unique responsibilities of women and men for protecting the built environment of the household were also the ideas to which agents of diffusion had attached the sale of gas and electric equipment in the first place.

BIBLIOGRAPHIC ESSAY

Sources on which I relied directly are cited in the notes. In this essay, I highlight the primary materials that proved valuable in allowing me to understand gas and electric operations in their urban and national contexts. I also focus on the books and articles that served as points of departure for several of my themes.

Two groups of business records proved especially valuable. In the basement of the former records center of the Public Service Company of Colorado (PSCC) in Denver, I noticed several large boxes with the name "Roy G. Munroe" written across the outside. Only because Roy Munroe's granddaughter had prepared a "scrapbook" and deposited it at the Denver Public Library was I aware that Munroe had worked for the company as a sales manager. In an age before public history and corporate archives existed, Munroe served as the company's "institutional memory." He had kept memorandums, reports, notes, newsletters, and minutes of sales meetings, all the "stuff" of mid-level management in a national corporation. Even more, Munroe, as the company's "historian," had prepared memos for other executives, who wanted to know more about the early days in appliances sales.

Much like Munroe, J. C. Nichols too relished his role as informal historian. Portions of his records at the University of Missouri at Kansas City consist of interviews and lectures in which he recounts his experiences in building the Country Club District and Plaza and the "lessons" acquired. Nichols's materials are the creation of a busy developer and civic leader who was conscious of himself as a leader among elite builders nationwide who were attempting to reshape land-use practices, shopping habits, and the interior environment of stores and homes. For real-estate practices in Denver, I relied on the *University of Denver Reports.*

In addition to business records, several bodies of trade literature informed the conceptualization of this book. In publications such as *Electrical World* and *Gas Age Record,* persons involved in the sales, finance, and organizational dimensions of gas and electric service reported on perceived problems and their solutions. Senior managers also permitted mid-level executives, such as Roy G. Munroe, to write articles for trade journals. Although the emphasis in these articles was the solution of problems associated with production, distribution, and marketing, one must be struck by the celebration of the core values that the writers believed had guided them to those solutions. Particularly valuable was the repeated invocation of the familiar symbols of their industry, such as the ideal of public service or the linear advance of engineering practice.

Directors of gas and electric firms, like their counterparts in other large corporations, also prepared newsletters for their employees. Executives of

Henry L. Doherty's companies published *Gas Service*, and their counterparts at the Kansas City Power & Light Company produced *The Tie* and *Power & Light News*. In each publication, managers connected such abstruse ideas as progress and public service to particular employees and to the details of daily operations, such as another sales campaign, new billing procedures, upgraded distribution equipment, or extensions of electric and gas lines farther afield.

Educators and other agents of diffusion produced their own trade literature. In publications like *American School Board Journal*, educators spoke of pedagogic methods, learning, and deportment as they discussed the virtues of heating and lighting in one period and air conditioning in the next. J. C. Nichols's employees and homeowners received the *Country Club District Bulletin*, which linked the development of new subdivisions with reports of technical improvements and idyllic images.

Early chapters of this book are rooted in a large body of scholarly literature focused on the political economy of the nineteenth-century city. For the entrepreneur and politician as partners during this period, the standard work is Charles N. Glaab, *Kansas City and the Railroads: Community Policy in the Growth of a Regional Metropolis* (Madison: State Historical Society of Wisconsin, 1962). More recent and equally useful analyses include Gunther P. Barth, *Instant Cities: Urbanization and the Rise of San Francisco and Denver* (New York: Oxford University Press, 1975); William Cronon, *Nature's Metropolis: Chicago and the Great West* (New York: W. W. Norton & Company, 1991); and Robin L. Einhorn, *Property Rules: Political Economy in Chicago, 1833–1872* (Chicago: University of Chicago Press, 1991). Books and articles dealing with the political economy of the nineteenth-century city have taken the place of an older body of literature focused on reformers or on the political boss and his cronies. Still valuable in that genre, however, is the classic essay by Samuel P. Hays, "The Politics of Reform in Municipal Government in the Progressive Era," *Pacific Northwest Quarterly* 55 (October 1964), 157–169, where the emphasis is on the ecological dimensions of urban politics.

More recently, historians have studied the city and its politicians and corporations as providers of vital services—transportation, water, electricity, and gas. Jon C. Teaford signaled the transition to this area of study in his "Finis for Tweed and Steffens: Rewriting the History of Urban Rule," *Reviews in American History* 10 (December 1982), 133–149. Teaford also demonstrates the remarkable willingness of urban politicians to invest in their city's infrastructure in his *Unheralded Triumph: City Government in America, 1870–1900* (Baltimore: Johns Hopkins University Press, 1984). On the other hand, Terrence J. McDonald, *The Parameters of Urban Fiscal Policy: Socioeconomic Change and Political Culture in San Francisco, 1860–1906* (Berkeley and Los Angeles: University of California Press, 1986), determines that a group of tightfisted politicians limited spending for improvements. Finally, however, Stanley K. Schultz, *Constructing Urban Culture: American Cities and City Planning, 1800–1920* (Philadelphia: Temple University Press, 1989), identifies a group as diverse as novelists and engineers who shaped a vision of the technologically oriented city and who went about the business of implementing it. Christine Meisner Rosen, *The Limits of Power: Great Fires and the Process of City Growth in America* (New York: Cambridge University Press, 1986), points out that proponents of infrastructural improvements still had to promote their projects through local politics structured to favor the proponents of inertia. For the provision of urban services and creation of the built environment since 1900, Roy Lubove, *Twentieth-Century Pittsburgh: Government, Business, and Environmental Change* (Pittsburgh: University of Pittsburgh Press, 1969), remains valuable.

Beginning around 1900, changing ideas about securing improvements in health and hygiene added to the attractiveness of gas and electric service. Books and articles by John C. Burnham, including *How Superstition Won and Science Lost: Popularizing Science and Health in the United States* (New Brunswick, N.J.: Rutgers University Press, 1987), alerted me to the importance of the physicians and scientists whose publications and lectures encouraged Americans to identify electricity with the possibility of achieving a technical fix for previously insoluble problems of public health and personal well-being. Near the conclusion of my research and writing, I was able to identify one of the foundations for the popular connection between the quest for healthful living and gas and electric service after reading Nancy Tomes, "The Private Side of Public Health: Sanitary Science, Domestic Hygiene, and the Germ Theory, 1870–1900," *Bulletin of the History of Medicine* 64 (Winter 1990), 509–539.

The cities for which these physicians, business leaders, and politicians were providing health and infrastructural services were never static. Urban populations increased rapidly in size and sorted themselves according to ethnicity and income. During the 1920s and 1930s, sociologists at the University of Chicago framed the questions and constructs that continue to inform research about the process they identified as residential succession. During the early 1960s, however, Sam Bass Warner's *Streetcar Suburbs: The Process of Growth in Boston, 1870–1900* (Cambridge: Harvard University Press, 1962), emerged as the standard and still valuable source for transportation and population deconcentration. Mark S. Foster, "The Model-T, the Hard Sell, and Los Angeles's Urban Growth: The Decentralization of Los Angeles During the 1920s," *Pacific Historical Review* 44 (November 1975), 459–484, extends Warner's findings to the automobile. Fred Viehe put industry ahead of transportation in accounting for deconcentration in his "Black Gold Suburbs: The Influence of the Extractive Industry on the Suburbanization of Los Angeles, 1890–1930," *Journal of Urban History* 8 (November 1981), 3–26. Toward the conclusion of the 1970s, historians and sociologists associated with the Philadelphia Social History Project brought another round of talent and sophistication to this area of inquiry. As a point of departure, I still like Eugene P. Erickson and William L. Yancey, "Work and Residence in Industrial Philadelphia," *Journal of Urban History* 5 (February 1979), 147–182.

After 1970, many urban historians turned their attention to the first generations of suburban residents and to the institutional arrangements they shaped. I found my thinking on the subject of elite builders and their customers influenced by Robert C. Twombly, "Saving the Family: Middle Class Attraction to Wright's Prairie House, 1901–1909," *American Quarterly* 27 (March 1975), 57–72; and by Mary Corbin Sies, "The City Transformed: Nature, Technology, and the Suburban Ideal, 1877–1917," *Journal of Urban History* 14 (November 1987), 81–111, particularly her focus on members of a group she identifies as the professional-managerial stratum. Margaret Marsh, *Suburban Lives* (New Brunswick, N.J.: Rutgers University Press, 1990), is an important treatment of the householders, and especially male householders, in the nation's new suburbs at the turn of the century. Marc Weiss, *The Rise of the Community Builders: The American Real Estate Industry and Urban Land Planning* (New York: Columbia University Press, 1987), explores the practices developed by elite builders and their subsequent embedding in federal policy. Finally, Michael H. Ebner, *Creating Chicago's North Shore: A Suburban History* (Chicago: University of Chicago Press, 1988), provides an overview of the planning and development of a suburban region and the subsequent institutional arrangements of several generations of upper-income residents.

Historians and sociologists have created an impressive body of literature treating business organizations in terms of responses to the changing "texture" of their environments, including their political environments. Originally, my thinking in this area was influenced by F. E. Emery and E. L. Trist, "The Causal Texture of Organizational Environments," *Human Relations* 18 (February 1965), 21–31. Thereafter, I found Louis Galambos and Joseph Pratt, *The Rise of the Corporate Commonwealth: United States Business and Public Policy in the Twentieth Century* (New York: Basic Books, 1988), an insightful overview of a large subject. In addition, Charles W. Cheape, *Moving the Masses: Urban Public Transit in New York, Boston, and Philadelphia, 1880–1912* (Cambridge: Harvard University Press, 1980), identifies the local and urban orientation of corporate behavior. Joel A. Tarr, "Railroad Smoke Control: The Regulation of a Mobile Pollution Source," in George H. Daniels and Mark H. Rose, eds., *Energy and Transport: Historical Perspectives on Policy Issues* (Beverly Hills, Calif.: Sage Publications, 1982), pp. 71–92, highlights the importance of urban public policy in forcing corporate leaders and members of their engineering staffs to revamp technical systems.

Portions of this book also revolve around the planning, administration, and expertise that characterized operations at the nation's gas and electric companies. Alfred D. Chandler Jr., *The Visible Hand: The Managerial Revolution in American Business* (Cambridge: Belknap Press of Harvard University Press, 1977), remains the standard source for these tendencies in American industry as a whole. For an understanding of organizational developments in the electrical industry, however, I relied on Thomas P. Hughes, *Networks of Power: Electrification in Western Society, 1880–1930* (Baltimore: Johns Hopkins University Press, 1983). Harold L. Platt, *The Electric City: Energy and the Growth of the Chicago Area, 1880–1930* (Chicago: University of Chicago Press, 1991), emphasizes the importance of political culture in establishing the framework within which utility executives such as Samuel Insull built a nationwide organization.

Ultimately, I determine that executives of gas and electric firms made choices not only as part of large organizations informed by a political culture, but also in terms of the social and political geography of cities. In addition to books and essays by Chicago sociologists, my initial thoughts in this area were shaped by Zane L. Miller, *Boss Cox's Cincinnati: Urban Politics in the Progressive Era* (New York: Oxford University Press, 1968), where the argument is that political outcomes were often a matter of the social ecology of the city. Similar and more recent studies to which I found myself referring included Christopher Armstrong and H. V. Nelles, *Monopoly's Moment: The Organization and Regulation of Canadian Utilities, 1830–1930* (Philadelphia: Temple University Press, 1986); Paul Barrett, *The Automobile and Urban Transit: The Formation of Public Policy in Chicago, 1900–1930* (Philadelphia: Temple University Press, 1983); and Harold L. Platt, *City Building in the New South: The Growth of Public*

Services in Houston, Texas, 1830–1915 (Philadelphia: Temple University Press, 1983).

Finally, historians have begun to study gas and electric service in terms of the enthusiastic reception they received among ordinary citizens and the meaning of that enthusiasm for shaping patterns of technological and cultural change. I returned frequently to the images and metaphors in Carolyn Marvin, *When Old Technologies Were New: Thinking About Electric Communication in the Late Nineteenth Century* (New York: Oxford University Press, 1988); in David E. Nye, *Electrifying America: Social Meanings of a New Technology, 1880–1940* (Cambridge: MIT Press, 1990); and in Mark J. Bouman, "Luxury and Control: The Urbanity of Street Lighting in Nineteenth-Century Cities," *Journal of Urban History* 14 (November 1987), 7–37.

The meaning of lighting for urban residents is part of the vast body of literature assessing the relationships among what scholars and popular writers deem technological and cultural elements. As I contend in the Epilogue, the earliest scholars of the subject, such as William F. Ogburn, assumed that technology and the engineers who dominated manufacturing corporations were able to influence social relationships—family relationships, for instance. More often cited in this area is Robert S. Lynd and Helen Merrell Lynd, *Middletown: A Study in American Culture* (New York: Harcourt Brace Jovanovich, 1927, 1957). By the 1970s, a new group of scholars identified politics, economics, or gender as the dominant factor. The design and manufacture of artifacts, and patterns of their consumption, flowed out of these institutionally defined arrangements, rather than the other way around, as Ogburn had contended. Early on, I was greatly impressed with the insights in George H. Daniels's classic essay "The Big Questions in the History of American Technology," *Technology and Culture* 11 (January 1970), 1–21. For culture and technology in urban settings, see Brian J. L. Berry, *Comparative Urbanization: Divergent Paths in the Twentieth Century* (New York: St. Martin's Press, 1981), first published in 1973 as *The Human Consequences of Urbanization.* As this study turned in the direction of householders, I encountered Charles A. Thrall, "The Conservative Use of Modern Household Technology," *Technology and Culture* 23 (April 1982), 175–194.

By the mid-1980s, a group of sociologists and historians calling themselves social constructivists had taken the leading positions in this debate about the relationships between culture and technology. The best introductions to that body of literature are several of the essays contained in Wiebe E. Bijker, Thomas P. Hughes, and Trevor Pinch, eds., *The Social Construction of Technological Systems: New Directions in the Sociology and History of Technology* (Cambridge: MIT Press, 1987). Finally, see Claude S. Fischer, *America Calling: A Social History of the Telephone to 1940* (Berkeley and Los Angeles: University of California Press, 1992), for a superb account of women's appropriation of the telephone in order to widen and deepen relationships.

Books and articles by six scholars in particular helped me shape the framework for this book. For the American corporation, I joined virtually every scholar in relying on Chandler, *The Visible Hand;* and Ruth Schwartz Cowan, *More Work for Mother: The Ironies of Household Technology from the Open Hearth to the Microwave* (New York: Basic Books, 1983), influenced my work in the area of gender and technology. Because of their focus on space and on the city as determinative rather than dependent factors, essays by Theodore

Hershberg and Stephanie Greenberg in Theodore Hershberg, ed., *Philadelphia: Work, Space, Family, and Group Experience in the Nineteenth Century* (New York: Oxford University Press, 1981), were important in my thinking. I remain greatly impressed with Thomas Hughes, *Networks of Power: Electrification in Western Society, 1880–1930* (Baltimore: Johns Hopkins University Press, 1983), especially for his conceptualization of systems of electrical, economic, and political power. For booming cities and suburbs and the importance of ideas, politics, culture, and transportation, I turned frequently to Kenneth T. Jackson, *Crabgrass Frontier: The Suburbanization of the United States* (New York: Oxford University Press, 1985). Finally, I learned much about politics, culture, cities, and infrastructure from each of many articles by Joel A. Tarr. One of my favorite is his "Changing Fuel Use Behavior and Energy Transitions: The Pittsburgh Smoke Control Movement, 1940–1950," *Journal of Social History* 14 (Summer 1981), 561–588.

Roland Marchand, *Advertising the American Dream: Making Way for Modernity, 1920–1940* (Berkeley and Los Angeles: University of California Press, 1985), points out: "It would be a wise author indeed who could pinpoint the sources of the underlying assumption and broad theoretical frameworks that have influenced his interpretation" (p. 426). This book and my own work are no exception. Nonetheless, I can say with some precision that I sought to fashion a monograph that integrated several bodies of scholarship. During my undergraduate years, the writings of Karl Marx, C. Wright Mills, Max Weber, and Louis Wirth (and several other Chicago sociologists) informed my understanding of politics, urban form, and organizational behavior. I went through graduate school during a historiographical moment emphasizing modernization, particularly the role of national business and professional elites in fostering the national and "universalistic" outlooks judged central to creation of a modern society. Most important for me among those authors and books were Samuel P. Hays, *Conservation and the Gospel of Efficiency* (Cambridge: Harvard University Press, 1959); K. Austin Kerr, *American Railroad Politics, 1914–1920: Rates, Wages, and Efficiency* (Pittsburgh: University of Pittsburgh Press, 1968); and Robert H. Wiebe, *Search for Order, 1877–1920* (New York: Hill & Wang, 1967).

By the late 1960s, scholars began to historicize the behavior of elites within specific locales and firms. One result of this scholarship was the rediscovery that elites were not always agents of modernization and that family and community had mattered greatly in the lives of most Americans, including many judged elite. My points of departure for this body of literature included Michael P. Rogin, *The Intellectuals and McCarthy: The Radical Specter* (Cambridge: MIT Press, 1967); John N. Ingham, "The American Urban Upper Class: Cosmopolitans or Locals?" *Journal of Urban History* 2 (November 1975), 67–87; and a first-rate textbook (supplemented by countless hours of conversation) by John G. Clark et al., *Three Generations in Twentieth Century America: Family, Community, and Nation* (Homewood, Ill.: Dorsey Press, 1977).

During the 1980s and early 1990s, localism and antimodernity emerged as two important themes in American historical scholarship. In particular, historical scholars (and others) were contending that agency was widely dispersed through American society. Occasionally, the ability to shape a portion of one's destiny was actually in the hands of ordinary persons—even the inarticulate, who were, it was argued, invoking local institutions and resources to preserve or advance their own interests. Early in the decade, Clifford Geertz, *Local Knowledge: Further Essays in Interpretive Anthropology* (New York: Basic Books, 1983), served as the conceptual sourcebook for many of these studies. The books and essays that informed my own work included James Borchert, *Alley Life in Washington: Family, Community, and Folklife in the City, 1850–1970* (Urbana: University of Illinois Press, 1980); Robert A. Slayton, *Back of the Yards: The Making of a Local Democracy* (Chicago: University of Chicago Press, 1986); Raymond A. Mohl, "Trouble in Paradise: Race and Housing in Miami During the New Deal Era," *Prologue: Journal of the National Archives* 19 (Spring 1987), 7–21; and Kenneth W. Goings and Gerald Smith, "'The Duty of the Hour': African-American Communities in Memphis, 1880–1920," a paper read at the Annual Meeting of the Organization of American Historians, Anaheim, California, April 1993.

Beginning late in the 1980s, Paul Barrett, Bruce E. Seely, and I explored the diminution of professional expertise and the apparent triumph in many arenas of locally oriented politics. Devolution of authority from the national to the local levels, and the disjuncture of economic and political institutions, were among the concepts we were exploring. In turn, our work was informed by books such as Stephen Toulmin, *Cosmopolis: The Hidden Agenda of Modernity* (New York: The Free Press, 1990); and Richard F. Hirsh, *Technology and Transformation in the American Electric Utility Industry* (New York: Cambridge University Press, 1989).

INDEX